Basic Electricity

Second edition

W. M. Gibson

Longman
London and New York

Longman Group Limited, London
Associated companies, branches and representatives
throughout the world
Published in the United States of America
by Longman Inc., New York

© W. M. Gibson 1969
Second edition © Longman Group Limited 1976

All rights reserved. No part of this publication may be
reproduced, stored in a retrieval system, or transmitted
in any form or by any means, electronic, mechanical,
photocopying, recording, or otherwise, without the
prior permission of the Copyright owner.

First published by Penguin Books Ltd., 1969
Second edition published by Longman Group Ltd., 1976

Library of Congress Cataloging in Publication Data

Gibson, W M
Basic electricity.

Bibliography: p.
Includes index.
1. Electricity. I Title.
QC522.G5 1976 537 76-7398
ISBN 0-582-44181-1

Printed in Great Britain by
Lowe & Brydone Printers Limited,
Thetford, Norfolk

UNIVERSITY OF BRADFORD
LIBRARY

040/538 539 26 JAN 1977

ACCESSION No.	CLASS No.
191320	M537 GIB
LOCATION	J. B. Priestley Library

To my family and colleagues, who have been generous
in encouragement, instruction and toleration.

Contents

Part Two Circuits and currents

Preface

This book is intended for the undergraduate student of physics in either the first or second year of a degree course and may be used on either of two levels:

(i) as a collection of important facts about electricity and magnetism expressed in conventional terms, with orthodox explanations of how they are related, or

(ii) as an introduction to the logical unity of electromagnetism, emphasizing some of the less-familiar links between well-known facts.

If it proves useful on the first level, it may also plant some seeds of interest in the minds of those who did not intend to use it on the second. It is written on the basis of interests and attitudes developed by research into the physics of elementary particles. The better known of these particles are the stuff of which ordinary matter is made. They are the carriers of electric current. In a very intimate sense, the electrons *are* electric charge. Electromagnetic fields and forces ought therefore to be describable in terms of the charges, positions and velocities of the particles giving rise to them. Although it is not necessary on every occasion to go right down to the individual source-particles to describe an electrical phenomenon, it ought to be possible to do so, or at least to see how one would do so if one wanted to. The present book sets out to show that this is not only possible, but illuminating; the whole of conventional electricity and magnetism is developed from the Coulomb law of electrostatic force, with invariance of electric charge under relativistic transformation.

To a particle physicist, this seems the nearest we can get to an ultimate description of what is going on. A field theorist may feel that the fields are the ultimate reality, or an engineer the currents, but it is no part of physics to claim that one valid description is better than another. However, it is emphatically part of a training in physics to learn that facts are describable on many different levels and that one fact may be related to another by so many chains of argument that neither can be called the 'consequence' or the 'cause'.

In electromagnetism there exists a particularly closely-woven mesh of interrelated facts and phenomena and it is possible to enter this mesh at many points, following linear chains of argument through it towards the more advanced parts of the subject. This is a useful and necessary exercise, but in carrying it out one must guard against the tendency to believe that there is something especially logical about the line chosen or something especially fundamental about one's starting-point. In particle physics we have begun to speak of a democracy of particles; of families in which no particle is more fundamental than another, none made up from the others, but each made necessary by the others. Corresponding to this in electromagnetism we may speak of a democracy of facts – all made mutually necessary by a web of logic. The beauty of the subject lies in the multiplicity and the symmetry of the threads in this web. Among the facts it is possible to pin to whichever we like the label 'foundation of the structure' We are now sufficiently adult to know that there is no profit in arguing the relative merits of historical, experimental or supposed logical foundations; each of these has already played a vital part in the development of the subject.

This book presents the system which starts with electrostatic force as the foundation of electricity, extends the discussion to moving charges by means of elementary relativistic mechanics and arrives at a description of magnetism and electromagnetism as inescapably associated with the electric forces between moving charges. This description is developed in the notation of conventional electromagnetic theory as far as Maxwell's equations, with a brief derivation of the wave equation for electromagnetic radiation and comment on the identity of the velocity given by it with the velocity used in relativistic transformation. The self-consistent circular argument relating these velocities with each other and with the constants needed in electromagnetic theory is a good example of the web-like nature of the subject. We retain the possibility of entering the logical circle at any point, moving round it in either direction, branching off it into one of the related fields, or ignoring it altogether while dealing simple-mindedly with a limited topic.

Within the ground covered by the present book, the student is encouraged to follow up lines of argument which have not been mapped out for him. He must not attribute any special authority to the limited selection which has been made here, with an eye to instruction and interest as well as to importance. Many topics have been mentioned in a deliberately incomplete or even provocative way, in order to stimulate thought and self-instruction.

SI units are used in this book, not through any conviction that they are best, but originally in order to achieve maximum usefulness while demonstrating that any self-consistent system will suffice. Now, in the second edition, the motivation for a more complete concentration on SI has been to ease the task of the student brought up on SI; many such students benefit more from uninterrupted concentration on one system than from interesting digressions into other systems. For the efficient deployment of effort, this strategy is effective up to a certain point. In advanced study beyond the scope of this book, the merits of systems other than SI, especially the Gaussian cgs system, continue to make their use desirable in certain texts, and in thinking about certain branches of the subject. Therefore an appendix is provided to translate the conclusions of this book into Gaussian cgs format, thereby (hopefully) easing the task of the student who needs to make the transition, e.g. to read an advanced text. Unnecessary proliferation of comment on other systems has been eliminated.

Further changes made in the second edition include (i) change to the marginally preferred convention that magnetisation should have the dimensions of $\mathbf{H}$ rather than of $\mathbf{B}$; and magnetic moment those of IA rather than $\mu_0 IA$; (ii) more careful discussion of the direction of fields in an electrically polarised material; (iii) distinction between Gauss's Law and Gauss's Theorem; (iv) introduction of Stokes's Theorem in an appendix; (v) improvement of the section on A.C.

bridges and (vi) insertion of many minor improvements and corrections.

My thanks go to Messrs. Longman for undertaking the publication of this second edition, to Penguin Ltd., and to Professor N. Feather who were responsible, as publisher and editor respectively, for the planning and publication of the first edition, to many friends and colleagues for help in planning, writing and revising this book, to Miss M. James and all who have contributed to necessary typing, and to my wife for understanding the abstraction that accompanies writing.

Part One
Fundamentals

Chapter 1
Electrostatic forces

1·1 **The nature of matter**

1·1 Nowadays every schoolboy can tell you that matter is made up of atoms, and can laugh at the (probably apocryphal) story of the motor-car that was advertised, in the days following the destruction of Hiroshima, as being made entirely of atoms.

We may therefore start by stating the fact which crystallographers have now established in so much detail, especially by experiments on the diffraction of X-rays: that solid matter is composed of atoms of dimensions of order 10^{-10}m. arranged in a more or less regular structure. In a perfect crystal of a salt the structure is a regular lattice, with atoms of different chemical elements occupying characteristic positions; the regularity of the lattice lies in the fact that the pattern of characteristic positions repeats itself at regular intervals along the axes of the crystal. Metals also may be obtained in crystalline form with regular lattices of identical atoms, but in the metals that we handle in everyday life regularity extends only over small regions which have escaped distortion in the mechanical processes of working the metal; they may have been displaced and rotated with respect to neighbouring regions but within themselves they have retained the structure characteristic of the crystalline state. In these circumstances, the physical properties of a metal depend upon the size of the regions of regular structure and on the presence in the lattice of occasional atoms of impurity elements.

1·2 The question which makes this book necessary is 'What is the force that holds atoms in their places in a solid?' Everyday experience convinces us that they are held rather firmly in their places, for solids are rigid and solid objects may be used to exert great forces on each other; they can withstand twisting forces, as in the coil-springs and torsion-bars used in the suspension of some motor vehicles, compressional forces like those on the bricks at the bottom of a building, and forces of extension as in a cable carrying a weight or in the metal of a cylinder containing compressed gas.

Searching among commonly experienced forces for a type that might hold atoms in their places we may be tempted to start thinking in terms of close packing of solid atoms. This can give us a model for a structure that can resist forces of compression; but to explain resistance to forces of extension or twist, an attraction between neighbouring atoms must also be considered. In any case, we have not really made much progress, for we have simply transferred the adjective in question, 'solid', from the matter as a whole to the individual atoms. It is still necessary to describe wherein lies the 'solidity' which prevents atoms from approaching infinitely close to each other.

For a description of the forces responsible for the mutual attraction of neighbouring atoms, as well as of such solidity as they possess, we must search our experience for forces which do not depend upon the properties of solids. Since these are what we are trying to explain, we must do it in terms of something else. What else is there, except gravitation? We may consider this as a candidate, until calculation shows us that it is much too weak. For example, gravitational forces can provide only about one part in 10^{29} of the tensile strength of a steel cable; this is about 10^{16} times too little to make it hold together as a solid in the neighbourhood of the Earth. Worse still, we find that if the atoms of a cricket ball were held together by gravitational forces only, rotation at the rate of one revolution per two hours would cause it to fly apart under centrifugal force.

1·1·3   So we must look for something much stronger than gravitation. One would not, at first consideration, expect to find the clue in the phenomena of static electricity; on the school lecture bench these amount to feeble efforts to overcome the force of gravity on little bits of paper by means of a fountain-pen which has been mysteriously rubbed. In everyday life we know static electricity as the source of the curious sensations that accompany the removal of a nylon garment on a dry day. Among the most noticeable of these sensations are crackling sounds and faint flashes of light; we are not aware that any forces are involved, until we examine the way the removed garment clings to the door against which it has been hung. We may hesitate between a physical and a physiological explanation for the strange appearance of the hair over which the garment has been dragged. However, it is worth thinking about the possibility that both may contribute, remembering that many a good idea has been lost through the semiconscious belief that phenomena have single causes.

1·2   Electric charge

Bodies can exist in two distinct types of electrically-charged state; the garments and hair of section 1·1 are not the most convenient experimental apparatus, but the meaning and validity of this statement can be checked by rubbing two rubber balloons with nylon cloths. For a short time afterwards (very short if the air is humid) the balloons will repel each other and the cloths will repel each other, but a cloth and a balloon will attract each other. Using the word electrical to describe these phenomena, we conclude from the observations that objects in similar electrical states repel each other, while an object in one type of electrical state attracts an object in the other. One of these two states is said to be positively charged and the other negatively charged.

Slightly more refined experiments with light spherical objects suspended on fine fibres serve to show that electric charges are quantities that can be shared, added and subtracted, if they are measured in terms of the forces which they exert on other charges. Let us suppose a body A is found to exert a force F on a test body T; then suppose the electric charge on A is shared between A and another body B by bringing them into contact. If A and B are placed successively at the original position of A, they now exert on T forces F_A and F_B which add up to a total F. Further, we find that positive and negative charges merit these names, since they are 'opposite' in two senses:

(i) two bodies which attract each other exert forces in opposite directions on a third charged body, when they are placed successively at the same position;
(ii) when two bodies which attract each other are allowed to touch, the respective forces which they exert on a third charged body are less than those which they originally exerted on it. This implies that their charges were opposite in the sense of partially cancelling each other when added.

At this level electric charge can be simply visualized as two sorts of fluid, of which at least one can be moved from place to place and which, since the two types attract each other, have a tendency to return to their normal, intermingled state which we call electrically neutral, or uncharged.

1·3　The electron

1·1　*The quantization of charge*

In order to probe more deeply into the nature of electric charge and the means by which it moves from place to place, we must refer to the results of some rather more elaborate experiments, notably those of Sir J. J. Thomson on electric discharges in gases at low pressure, carried out between 1895 and 1915. These and similar experiments showed the existence of electrically charged particles with well-defined ratios of charge to mass.

The description of these ratios in terms of a basic quantity of charge on particles of different masses, became firmly established when the magnitude of this basic charge was accurately measured by R. A. Millikan in 1917. In this experiment, which is frequently repeated in the teaching laboratories of schools and universities, droplets of oil are observed through a microscope as they are allowed to fall through the space between two horizontal metal plates. These plates can be electrically charged by means of a battery. If a drop happens to carry an electric charge, its velocity of fall will change when the battery is connected; it may even stop falling or start to rise if the electrostatic force on it is greater than the gravitational force due to its mass. When it is in motion its velocity depends upon the viscosity of the air around it as well as on its mass, radius and electric charge. These experiments led to the conclusion that all electric charges are made up of whole numbers of identical, indivisible, small charges. A fraction of this basic charge has never been unambiguously observed (see section 1.3.4), but positive and negative charges of one, two, three and more times the basic charge are found. When we have a very large charge, it is difficult to be sure which whole number of basic charges make it up and impossible to confirm that it is a whole number, but experiments like Millikan's can distinguish clearly between whole numbers up to twenty or more with sufficient resolution to show that fractional charges do not occur.

5　Electrostatic forces

In order to give electric charges to the oil-drops in Millikan's experiment, it is necessary to break some of the molecules of air into charged fragments which we call ions. This is often done by shining ultra-violet light or X-rays into the space between the plates. If after this an oil-drop captures one or more ions, it acquires the total charge. The natural inference from these observations is that atoms contain at least one type of positively charged particle, and at least one type of negatively charged particle; also that the charges on each type are identical and furthermore that the charges on opposite types are equal in magnitude so that normal atoms can be electrically neutral.

The experiments of J. J. Thomson, which were mentioned above, throw some light on the properties of these constituent parts of atoms; the observed ratios of charge to mass for particles in electric discharges in gases extend to much larger values for negatively charged particles than for positively charged. If all the charges are of similar magnitude, this means that there are negatively charged particles which have smaller masses than any of the positively charged particles. These light negatively charged particles are the electrons.

1·3·2 *The Bohr model of the atom*

For the arrangement of these constituent parts in normal atoms in contrast to their arrangement in the ions formed when atoms are broken up by ionizing radiation, or electric discharge, we turn to the Bohr model of the atom. This was established following the experiments of Rutherford and others around 1913, which showed that the positive electric charge is carried by a massive but very small nucleus at the centre of the atom. The negative charge is carried by particles of small mass, called electrons, which normally form the outer layers of each atom.

1·3·3 *The hydrogen atom*

The simplest atom, that of hydrogen, has one electron, and its nucleus is a positively charged particle which is given the special name of proton. Larger atoms have nuclei which behave as if they were made up of roughly equal numbers of protons and neutrons: the neutron is an electrically neutral particle having nearly the same mass as the proton. The number of protons in the nucleus of an atom is called its atomic number, and this is the same as the number of electrons present in the neutral state of the atom.

1·3·4 *Two problems: the balance of electronic and protonic charges; fractional charges*

It is worth mentioning here that the perfect balance between the charges of the electron and the proton, which has been shown to hold within a factor 4×10^{-20} by A. M. Hillas and T. E. Cranshaw (*Nature*, vol. 184, 1959, p. 892), is a serious problem in present-day theories of elementary particles. Nuclei have properties which are determined almost entirely by the strong forces which hold their constituent protons and neutrons together; whereas electrons have hardly any properties except electromagnetic ones. Insofar as electrons and protons have structure, they do not appear to have enough in common to ensure their having equal and opposite charge.

This puzzle has been exacerbated by the discovery that protons and other strongly interacting particles behave as if they were made up of fractionally charged particles. These hypothetical particles, which have come to be known as quarks, would if observed singly have charge one third or two thirds that of the electron. Experimental searches for these particles, while yielding occasional results that could be attributed to free quarks, have so far yielded none that are incapable of explanation in terms of ordinary particles. It is thus not known whether the quarks are real particles capable of independent existence, or unphysical elements in a convenient model of a mathematical theory. Thus the statements of section 1.3.1. about the nonexistence of fractional charges remains valid for all normal circumstances, and at present for all known circumstances.

3·5   The value of the electronic charge

A positive ion will normally be a neutral atom stripped of one or more electrons. Of the negative ions, the lightest are free electrons while the more massive are ordinary atoms which have acquired one or more extra electrons.

The charges on these ions are multiples of a basic charge which is given in terms of the coulomb (a unit of charge which we shall define in Chapter 6) as:

$$e = 1 \cdot 60207 \times 10^{-19} \text{ coulombs}$$

(Millikan's experiment has been checked and corrected by more advanced work.) As the charge of the electron is negative it must be written as:

$$-e = -1 \cdot 60207 \times 10^{-19} \text{ coulombs}$$

This figure is quoted as being accurate to within a probable error of forty-five parts per million.

1·4   The structure of solids

4·1   The structure of salts

We are now in a position to give a qualitative answer to the question posed in section 1·1. Solids of the type known as salts contain atoms which are permanently ionized, some positively charged by having lost an electron and an equal number negatively charged by having gained an extra electron. If some atoms are doubly charged through having lost or gained two electrons, the numbers will not be equal, but will be such as to make the structure as a whole electrically neutral. The oppositely charged ions exert strong forces of attraction on each other; these forces are balanced by the mutual repulsion of the outer layers of electrons, which prevents interpenetration of atoms and makes them behave as if they were solid spheres with definite radii. Attempts to stretch or distort a structure of this nature lead to slight increases in the distances between some pairs of ions. In these circumstances part of the electrostatic attraction between the ions must be balanced by the externally-applied force.

1·4·2   *The structure of metals*

In metals, however, we do not have two types of atom with a tendency to become oppositely charged ions by transfer of electrons, but one type of atom with a common tendency to release electrons into a continuum of mobile electrons which have given up individual association with particular atoms. The resultant array of massive positive ions in a sea of negatively charged electrons forms a stable structure on roughly the same scale as that which holds for crystalline salts. In fact, pure metals solidifying slowly under suitable conditions do so in a form which has the regularity characteristic of the crystalline state. A perfect crystal of a pure metal differs from one of a salt in that it is extremely weak. This is because the planes of positive ions can slide rather freely over each other. So long as the planes are perfect, sliding occurs easily, and the structure has little physical strength; but when the metal has been twisted and worked enough to break up any initial large-scale regularity, it is much stronger. A similar effect may be achieved by alloying several metals together; the ions of different sizes can interlock successive planes firmly enough to make a rigid structure.

In Chapter 3 we shall discuss how the mobile electrons in a metal are responsible for the conduction of electricity.

1·5   The inverse square law

Simple experiments on the forces between two electrically charged objects show approximately that the forces between them are inversely proportional to the square of their distance apart. This provides one similarity between electrostatic forces and the enormously weaker forces due to gravitation. Direct confirmation that electrostatic forces follow an inverse square law is necessarily rather rough, but indirect evidence shows that the law holds to within one part in 10^{10}. In other words, no departure from the inverse square law has ever been observed and there are strong theoretical reasons for expecting this situation to continue.

Chapter 2
The electric field

2·1　The magnitude of electrostatic forces

Following the discussion of electric charge in Chapter 1, we may write the force exerted on a charge q by a charge Q at a distance r in vacuum in the form:

$$\text{Force} = \frac{1}{4\pi\varepsilon_0}\frac{qQ}{r^2} \tag{2.1}$$

The constant in this equation is written as $\dfrac{1}{4\pi\varepsilon_0}$ in order to bring our formulae into line with the conventions of the Système International (SI). In this system distances are measured in metres, masses in kilogram and times in seconds, so that the unit of force is the kg.m.sec.$^{-2}$ or *joule per metre*, which is called the newton. Since charges are measured in coulombs, equation (2.1) requires ε_0 to have the numerical value $8\cdot85\times10^{-12}$ (in units which may be called coulomb^2 newton^{-1} m.$^{-2}$, or [see page 54] farad m.$^{-1}$). For the time being this may be taken as experimental fact, though we shall see in Chapter 6 that it is a necessary consequence of the way we use electromagnetic effects to define the ampere, the unit of electric current. We then define the coulomb as the charge carried by a current of one ampere flowing for one second and the value of ε_0 follows.

The electrostatic force on q is directed along the line drawn from Q to q, and like any other type of force it may be written as a vector; the vector form of equation (2.1) is:

$$\mathbf{F} = \frac{1}{4\pi\varepsilon_0}\frac{qQ}{r^2}\frac{\mathbf{r}}{r} \tag{2.2}$$

$\mathbf{r}$ is the vector specifying the position of q with respect to Q, and r is its length, so that $\dfrac{\mathbf{r}}{r}$ is a unit vector.

It is interesting to note at this point that equations (2.1) and (2.2) imply very large forces. Thus if it were possible to store on a human body a charge of one coulomb, sufficient to light a torch bulb for 3 seconds, two such bodies would repel each other with a force of 100 tons weight at a distance of 100 metres (or attract each other if the charges were of opposite signs). If the bodies were brought to within one yard of each other, the force would be equivalent to a weight of a million tons. Forces even more astronomical in magnitude would be obtained if it were possible to remove all the electrons from the bodies, since this would leave them with net positive charges of order one thousand million coulombs.

2·2 The nature of electric field

At one time it seemed meaningful and important to argue whether a charge Q exerted an electrostatic force on another charge q by action at a distance, or by creating in the neighbourhood of q a field which exerted a force on any charge placed there. To many the idea of action at a distance was philosophically repugnant and the field seemed to provide an intermediate mechanism.

Nowadays, however, such argument is unnecessary since the two approaches are seen as being simply two ways of describing the same reality: the field is not a mechanism which exerts forces, but a verbal and mathematical device for saying that a test-charge will experience a force. When we want to discuss the force on a test-charge without specifying the detailed configuration of source-charges which is responsible for the force, we say that a force F on a charge q implies the existence at the position of q of an electric field E, according to the vector equation:

$$\mathbf{F} = q\mathbf{E} \tag{2.3}$$

In simple words, the electric field at a point is given in magnitude and direction by the force per unit charge experienced by a test-charge placed at that point. The sense of the field is that in which a positive test-charge would tend to move. If the force is due to a single source-charge Q at a distance r, equations (2.2) and (2.3) require that the field should be:

$$\mathbf{E} = \frac{1}{4\pi\varepsilon_0} \frac{Q}{r^2} \frac{\mathbf{r}}{r} \tag{2.4}$$

2·3 Potential

2·3·1 *Work done in a small displacement*

So far, the forces occurring between electric charges at fixed positions have been discussed; now we must consider the possibility of changes in these positions and the work which may be involved in such changes.

If an electrically charged object, experiencing a force F by virtue of its charge, is allowed to move a small distance δs in the direction of F, an amount of work $F\delta s$ is done on it. This work is provided by the electrostatic potential energy of the system, which decreases by $F\delta s$. The energy may reappear in one or more of three forms, according to the situation of the object: (a) if it is moving freely the energy will become kinetic energy of its motion; (b) if its movement is resisted by frictional or viscous forces, the energy will be dissipated as heat; (c) if the object is supported by conservative forces (like those provided by perfectly elastic springs) the movement may take place reversibly in such a way that the energy $F\delta s$ is transferred from the electrostatic potential energy to the mechanical potential energy of the system. When the reversible movement takes place in the other direction, with external forces causing the object to move a small distance δs in opposition to the force F, an amount of work $F\delta s$ is done by the external forces and the electrostatic potential energy of the system increases by $F\delta s$. For convenience we shall consider only reversible displacements.

3·2   Work done in a small displacement in a general direction

If the force and the displacement are at right-angles to each other no work is done. This allows us to treat displacements and forces which are in general directions by resolving them into components along three mutually perpendicular axes. We consider a vector displacement δs as being made up of displacements δx along the x-axis, δy along the y-axis and δz along the z-axis. If the force acting on the charge is F with components F_x, F_y and F_z along the three axes, the total work done by the field is:

$$\delta W = F_x\delta x + F_y\delta y + F_z\delta z \qquad (2.5)$$

The values of F_y and F_z do not affect the work done in the displacement δx, so terms like $F_y\delta x$ do not appear. The right-hand side of equation (2.5) is the quantity discussed in Appendix B as the scalar product of the two vectors F and δs. It is usually written $F.\delta s$ and is equal to $F\,\delta s\cos\theta$, where F is the magnitude of F, δs is the magnitude of δs and θ is the angle between them. This may be interpreted as δs multiplied by the projection of F in the direction of δs, or as F multiplied by the projection of the displacement in the direction of F.

3·3   Work done in a displacement along a path

If a charge q is moved through a large distance along a path s, the total work done by the field is the sum of the scalar products $F.\delta s$ for all the small elements δs which together make the path s. In the limit of infinitely small elements, this sum may be written as an integral:

$\int F.ds$ = total work done by the field

An integral of this type is called a line integral; it may be considered as an ordinary integral along the path of the quantity $F\cos\theta$, which is the component of F in the direction of each element of the path. Alternatively the line integral may be considered as the sum of three ordinary integrals $\int F_x\,dx$ along the x-axis, $\int F_y\,dy$ along the y-axis and $\int F_z\,dz$ along the z-axis.

Throughout all this we may replace F by qE, and obtain as an alternative expression for the total work done by the field:

$q\int E.ds$

Thus the quantity $\int \mathbf{E} \cdot \mathbf{ds}$ along a specified path represents the work done by the field per unit charge in moving a charge along it and $-\int \mathbf{E} \cdot \mathbf{ds}$ is the work done by external forces.

2·3·4 *The potential difference between two points*

We now consider two different paths from point A to point B. If the line integral $\int \mathbf{E} \cdot \mathbf{ds}$ had different values along the two paths, it would be possible to allow a charge to move with the field along the path with the larger value of $\mathbf{E} \cdot \mathbf{ds}$, and then make it move back to its original position along the other path without using up all the energy gained from the first path. Repetition of this process could be made the basis of a perpetual motion machine which would go on continuously extracting energy from the field. If however the field is electrostatic, i.e. due to a stationary source-charge, it cannot act as a source of energy indefinitely. Since perpetual motion machines do not exist, we infer that in an electrostatic field $\int \mathbf{E} \cdot \mathbf{ds}$ must have the same value for any path between two points. This value is characteristic of the two points, and $-\int \mathbf{E} \cdot \mathbf{ds}$ is called their electric potential difference.

The source-charge may be considered as setting up an electric potential which has a definite value (say V) at each point. This involves choosing a standard point for the zero of potential and using the term potential at a point when we mean potential difference between that point and the standard. The difference in potential of any two points then represents the work done in moving a unit charge between the two points. A positive test-charge tends to move away from a point at higher potential, so when a positive charge moves from a higher to a lower potential (or a negative charge from a lower to a higher potential) work is done by the field; conversely, work must be provided from outside to make a positive charge move from a lower to a higher potential (or a negative charge from a higher to a lower potential).

Potentials are measured in joules per coulomb, which are called volts. The unit of electric field is therefore the volt per metre.

2·4 Field due to a single point charge

It follows from equation **(2.4)** that the electric field of a single charge located at the origin is everywhere directed radially outward, and that its magnitude at a distance r is proportional to $\dfrac{1}{r^2}$. When considering the potential set up by such a charge we see that no work is done in moving a test-charge at right-angles to the line drawn to it from the origin. All points on the surface of a sphere with its centre at the origin are therefore at the same potential, or to make an equivalent statement with reversed emphasis, the equipotential surfaces are spheres with centres at the position of the source-charge.

The potential difference between points on two of these spheres may be calculated by taking the line integral $\int \mathbf{E} \cdot \mathbf{ds}$ between them, and since E is everywhere radial it is easiest to take the integral along a radial line between the two spheres. Substituting the magnitude of **E** from equation **(2.4)** we get for the potential difference between two points at radial distances r_2 and r_1:

$$V = -\int_{r_1}^{r_2} E \, dr$$

$$= -\frac{1}{4\pi\varepsilon_0} \int_{r_1}^{r_2} \frac{Q}{r^2} \, dr$$

$$= \frac{1}{4\pi\varepsilon_0} \left[\frac{Q}{r}\right]_{r_1}^{r_2}$$

$$= \frac{Q}{4\pi\varepsilon_0} \left(\frac{1}{r_2} - \frac{1}{r_1}\right) \tag{2.6}$$

In order to consider potentials as well as potential differences, as explained in section 2·3, a standard point whose potential is defined as being zero is needed. For the single point charge which we are now considering a convenient standard is provided by a hypothetical point at infinite distance. If in equation **(2.6)** we put r for r_2, and ∞ for r_1 we find that the potential at r with respect to the standard point at infinity, is:

$$V = \frac{Q}{4\pi\varepsilon_0 r} \tag{2.7}$$

It should be noted that an inverse square law of force leads to a potential inversely proportional to the first power of r.

2·5 Field lines and equipotential surfaces

For the especially simple case of a single point charge we have seen that the electric field, being radial, is everywhere perpendicular to the equipotential surfaces, which are concentric spheres.

In fact this relationship between the equipotential surfaces and the lines of electric field is not peculiar to the case of a single point charge. We shall now show by two arguments and one analogy that the electric field at a point must be perpendicular to the equipotential surface at the same point, even for configurations of source-charges which give much less simple shapes of equipotential surface. Figure 1 illustrates an example of such a configuration, the field near the edge of a parallel-plate condenser.

(i) Consider two equipotential surfaces very close together, as shown in figure 2. AB is drawn perpendicular to these surfaces, with A on one surface and B on the other; also AC is drawn not perpendicular to the surfaces, but with C on the same equipotential surface as B. AC being equal to $\sqrt{(AB^2 + BC^2)}$ must be longer than AB. However, since B and C are at the

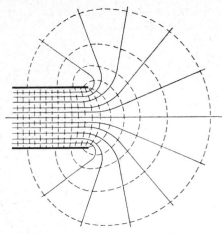

Figure 1. Equipotential surfaces (full lines) and electric field (broken lines) near the edge of a parallel-plate condenser.

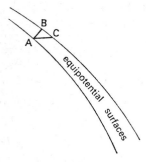

Figure 2. To show that the electric field is perpendicular to the equipotential surfaces (see text).

same potential, the potential difference between B and A is equal to that between C and A; each of these potential differences is equal to the length of the line (AB or AC) multiplied by the projection along it of the electric field E. Thus E must have a projection along AB larger than that along AC. Therefore E has a larger projection in the direction perpendicular to the equipotential surfaces than in any other direction and E itself must be a vector in this direction.

(ii) The second argument runs as follows: an equipotential surface is one in which small displacements of a test-charge involve no work. Any small displacement which is not perpendicular to the electric field E will have a component parallel to E and moving a test-charge through it will involve a non-zero amount of work. Therefore equipotential surfaces can only be perpendicular to **E**.

(iii) As a two-dimensional analogy we may consider the surface of a mountain on which height may be called gravitational potential. Equipotential surfaces become contour lines. The gravitational field, for objects restricted to the surface, is the gradient in the direction of the line of steepest descent. Experience with skis or with maps quickly convinces that this line is everywhere at right-angles to the contour lines, and also that the magnitude of the gradient is greatest where the contour lines are closest together, a fact which leads us to the treatment of potential gradient in the next section.

2·6 Field as potential gradient

We have already seen that electric field and potential are related in a special way: by introducing potential difference as a line integral of electric field along a path, we found that equipotential surfaces must everywhere be perpendicular to the electric field.

Alternatively, we may start by defining the potential at a point as the energy per unit charge required to bring a test-charge up to that point from an infinitely remote reference point. Then just as the work done is the integral of the force along a path, so the force in a given direction is the work done per unit distance for displacements in that direction; it follows that if the potential changes by an amount δV (i.e. the work done by the field is $-\delta V$), then as we make a small displacement δx parallel to the x-axis a test-charge will experience a force per unit charge (in the x-direction) equal to $-\dfrac{\delta V}{\delta x}$. In the limit of small displacements, $\dfrac{\delta V}{\delta x}$ becomes the differential coefficient $\dfrac{dV}{dx}$.

But in order to allow for the possibility of V depending on y and z as well as on x, we must use the partial differential coefficient $\dfrac{\partial V}{\partial x}$ which represents the rate of change of V with x when y and z are kept constant. Then the argument gives us the x-component of force per unit charge, i.e. the x-component of electric field. In this way, we see that the electric field **E** is a vector of which the x-, y- and z-components are the three (negative) partial differential coefficients $-\dfrac{\partial V}{\partial x}$, $-\dfrac{\partial V}{\partial y}$, and $-\dfrac{\partial V}{\partial z}$. This fact is summarized by saying that the electric field is the gradient (see Appendix B) of the potential, with the sign changed.

2·7 Superposition of fields

The electrostatic force exerted on a charge q by a source-charge Q is not affected by the presence of a second source-charge Q': the combined force on q due to Q

and Q' is just the resultant of the forces produced by Q and Q' separately. For source-charges at the same position, this fact has already been included in equation (2.4), where the force is made proportional to Q: this requires that the force due to the whole of a divisible charge Q is equal to the sum of the forces which would be produced by the parts of Q acting separately.

A simple way of describing this additive property of electrostatic forces is to say that the electric field of a set of source-charges is the same as that obtained by superposing the fields which the individual source-charges would have if they were acting alone. We perform the superposition of two fields by taking the resultant or vector sum of the two values of electric field E at each point; for consistency with this, we must also take the algebraic sum of the potentials.

This principle of superposition allows us to calculate the electrostatic effects of a distributed charge by adding the effects of small parts of it. In the limit of very small parts, assumed continuously divisible for convenience, this process becomes integration: it is used directly in Chapter 10.

The same principle is also invoked in Chapter 4 when we go from the effect of a single moving charge to that of many charges moving in an electric current. Since the effect of a moving charge is described in terms of an electric and a magnetic field, it follows that magnetic fields are subject to a law of superposition similar to that which applies to electrostatic fields.

In this book, as elsewhere, superposition will often be assumed to be valid without explicit mention of it; the difficulty of treating a subject without superposable fields need hardly be emphasized.

Chapter 3
The electric current

3·1 Conduction of electricity

3·1·1 *The model of a solid*

The familiar phenomena of the flow of electric currents may be understood in terms of the concepts discussed in Chapters 1 and 2. A full discussion of real conducting materials needs advanced quantum theory, so we shall use instead a simple descriptive model which is useful as a basis for qualitative discussion, and to give orders of magnitude, but should not be taken too literally.

We visualize a metal as a more-or-less regular lattice of fixed positively-charged ions surrounded by mobile, negatively-charged electrons. The mobile electrons are sometimes said to form a continuum; this is because their energies are continuously variable, unlike those of electrons attached to individual atoms, which are restricted to certain fixed values. The electrons of the continuum are in constant random motion colliding with the positive ions and sharing thermal energy with them. Thus at a given temperature, the thermal energy of a piece of metal has a fixed total value made up in definite proportions of energy of random motion of the electrons and energy of vibration of the positive ions. One can indeed imagine the specific heat to be made up of two parts: the specific heat of the lattice, which predominates at ordinary temperatures, and the specific heat of the electrons, which is important only at the lowest temperatures. These two contributions cannot be observed separately, since it is impossible to maintain the electrons and the lattice at different temperatures.

3·1·2 *Electron drift*

When the two ends of a piece of metal are at different electric potentials there is an electric field in the metal, under the influence of which the electrons experience a force and are accelerated in the direction opposite to the field (opposite because of the negative charge). However, the acceleration is quickly swamped in the random motion of the electrons and the energy gained is shared with the positive ions as collisions take place. The net effect is to superimpose on the random distribution of velocities of the electrons a drift velocity in the direction opposite to the field but of magnitude proportional to that of the field. The process may be compared with that in which a pressure gradient causes a viscous fluid to take up a constant velocity through a porous solid.

If there are N mobile electrons per unit volume and the drift velocity is v, then a total charge Nev passes through unit cross-section in unit time. Such a flow of charge is called an electric current and the rate of flow is measured in amperes, which are coulombs per second. Most metals have about one mobile electron per atom so that N is of order 7×10^{28} per m.^{3} If each electron carries a negative charge of 1.60×10^{-19} coulomb, a current of 1 ampere in a wire of cross-sectional area 1 mm.^{2} involves a drift velocity of:

$$v = \frac{\text{current}}{\text{area}} \times \frac{1}{Ne}$$

$$= 10^6 \times \frac{1}{(7 \times 10^{28} \times 1.60 \times 10^{-19})} \text{ m. sec.}^{-1}$$

$$= 10^{-4} \text{ m. sec.}^{-1}$$

This velocity is very low and the student is invited to comment upon the fact that operating a switch may produce effects which travel along a circuit at a very much greater velocity. (More fully-informed comment will be possible when the student has considered transmission lines.)

3·2 Resistance

The current in a metal wire may be denoted by:

$$I = nev = NAev$$

where A is the cross-sectional area of the wire, N the number of mobile electrons per unit volume, n the number of mobile electrons per unit length of wire, v the drift velocity and e the magnitude of the charge on each electron. We have seen that v is proportional to the electric field in the metal and therefore to $\dfrac{V}{l}$ if the field results from the application of a potential difference V to a wire of length l. If we put

$$v = k\frac{V}{l}$$

we get

$$I = kNe\frac{A}{l}V \qquad \qquad (3.1)$$

$kNe\dfrac{A}{l}$ is a product of factors all relating to the particular piece of metal (except e which is universal) and is called its conductance. The reciprocal of the conductance is called the resistance, and is usually given the symbol R. In the equivalent forms:

$$I = \frac{V}{R} \qquad V = IR \qquad\qquad\qquad (3.2)$$

equation **(3.1)** will be recognized as an expression of Ohm's law.

Resistances are measured in volts per ampere, which are called ohms. When the value of a resistance is being specified the word 'ohms' is often abbreviated to the single Greek letter Ω.

3·3   The status of Ohm's law

It is appropriate at this stage to comment upon the status of Ohm's law: it expresses a proportionality between the current in a conductor and the potential difference across it. We have seen that such a proportionality may be expected in the special case of conductors containing carriers of charge which reach a drift velocity proportional to the applied electric field. Metals and many other conducting materials also come into this category for most practical purposes, the proportionality being quite exact. However, many substances and devices show definite departures from this proportionality: for example, in semiconductors the number of mobile carriers of charge is not constant, while in electrolytic cells an extra potential difference is required to convert neutral atoms to charged ions at one electrode and vice versa at the other. In rectifiers of all types there is a greater supply of carriers to take charge in one direction than in the other, and in temperature-sensitive resistors the passage of current may raise the temperature, thereby changing the resistance and giving a relationship between current and potential difference, which is non-linear by an amount depending on the heat losses and on the time for which each current is maintained.

Thus Ohm's law does not express any universal truth of the type which is contained in the laws of conservation of energy and momentum; used in orthodox circumstances, it merely describes a convenient property of ordinary materials.

3·4   Resistivity

Returning to equation **(3.1)** we see that the resistance of a specimen depends on its dimensions, and may be written as:

$$R = \rho \frac{l}{A} \qquad\qquad\qquad (3.3)$$

where ρ contains all the factors which are independent of the dimensions ($\frac{1}{kNe}$ in our simple model), and is called the resistivity of the material.

According to equation **(3.3)** the resistance of a unit cube of the material, for current flowing parallel to one set of edges, is numerically equal to the resistivity. But for specimens of other shapes ρ must be expressed as the resistance of unit cross-sectional area per unit length, and is therefore measured in ohm-metres.

19   The electric current

3·5 Heating effect of an electric current

We have already mentioned that when a potential difference is causing a current to flow in a metal, the electrons acquire extra energy as well as extra velocity and thus as the electrons collide with the fixed positive ions, this becomes energy of random motion, that is to say, thermal energy. This extra thermal energy, or heat to give it its everyday name, serves to raise the temperature of the metal. The amount of heat generated may be calculated from our definition of potential difference: unit potential difference is that which imparts unit energy to a coulomb of charge moving under its influence. A charge Q, moving through a potential difference V, acquires an amount of energy QV. In a metal, this energy is not accumulated but acquired, a little at a time, as kinetic energy of the electrons, and converted into heat by the subsequent collisions. The amount of heat generated therefore may be described as QV joules, or alternatively as $\dfrac{QV}{J}$ calories, where J is the mechanical equivalent of heat expressed in joules per calorie. However, since heat is merely a special type of energy it is convenient for many purposes to think of it in the same units as are used for mechanical and electrical energy; for if we form the habit of expressing specific heats in joules per degree per unit mass, we may avoid use of the calorie altogether, even for calculating temperature changes.

When a current I is flowing across a potential difference V, I gives the charge passing per second and therefore IV gives the heat liberated in joules per second, which are called watts. From the law of conservation of energy IV must also be equal to the rate at which external agencies are supplying energy to the conductor. This energy is supplied by taking electrons at the point of higher potential, and returning them to the conductor at the point of lower potential: work has to be done on negatively-charged particles to move them from a point of higher to a point of lower potential. The necessary work can be obtained, at the expense of chemical energy, in an electrolytic cell or battery, which provides the simplest source of potential difference. In another example, it is obtained at the expense of gravitational energy, via a magnetic field, in a hydroelectric generating station. These and all other continuously-operating sources must involve forces which are not describable in purely electrostatic terms.

3·6 The electric circuit

From the foregoing discussions we may infer that a steady current will flow only when there is a complete circuit with a device that causes the same electrons (or electrons from an unchanging total) to circulate continuously in one direction. Since a definite potential difference is required to force the electrons through the passive part of the circuit at a given rate, the circulating device must be active in that it provides energy with which to create this potential difference. It is sometimes convenient to use the special name electromotive force (abbreviated to e.m.f.) for the potential difference created by an active device such as a cell or a dynamo.

The simplest electrical circuit therefore consists of a source of e.m.f. with two terminals which are connected through pieces of wire to the terminals of a passive component called the load. If the resistance of the load is much larger than that of the connecting wires, the potential difference between the ends of each wire can be neglected, and the potential difference across the load may be put equal to the e.m.f. of the source. A circuit such as this will demonstrate the most elementary fact of electrical circuitry; that a break at any point in the circuit interrupts the flow of current round it.

The flow of steady currents in simple circuits and in circuits with branches is governed by quantitative laws which are discussed in the following section.

3·7 Kirchhoff's laws

3·7·1 First law

If we have a circuit containing alternative paths for current, the distribution of current between the different paths is controlled by two laws each of which may be expressed in several forms.

The first is the law of conservation of electric charge, which merely describes the fact that electrons and ions carry fixed electric charges which cannot be destroyed in any way. They may be rendered temporarily less effective by combination with particles of opposite charge or more effective by association with large numbers of similarly-charged particles; but the arithmetic total of electric charge in a closed system cannot be varied. This holds even for the most violent chemical and nuclear reactions, as well as for ordinary processes. It means that when electric currents are flowing, the amount of charge per second reaching a region must be equal to the amount of charge per second leaving it, unless an electrostatic charge is being built up in it. Therefore, in a circuit carrying steady currents, the total current reaching a junction is equal to the total current leaving it; alternatively, if we attribute signs as well as magnitudes to currents, we may say that the algebraic sum of all the currents reaching a junction is zero. In this form the law of conservation of charge is known as Kirchhoff's first law.

3·7·2 Second law

The second law is that of conservation of energy which we have invoked many times in Chapter 2. When applied to an electric circuit this requires that the total work done on an electric charge travelling round a closed path must be zero. That is, the amount of work done on it by any active source of e.m.f. in the path must be equal to the amount of energy which it dissipates in going through the potential differences of all the passive parts of the path. In other words the total e.m.f. of any sources of electric power in the path must be equal to the total of the potential differences across the passive elements of the circuit.

This conclusion applies to any closed path in a circuit no matter how many side branches are attached to the path; it is the basic form of Kirchhoff's second law. A more familiar form is obtained when we insert the assumption that all the passive elements obey Ohm's law: in this case, the potential difference across each element is equal to the product of its resistance and the current through it. In the circuit shown in figure 3 the currents through the elements 1, 2, 3 and 4

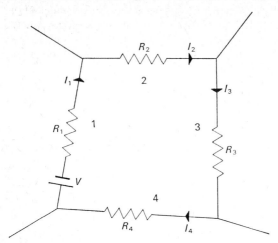

Figure 3.   Circuit to illustrate Kirchhoff's second law.

are different because of the currents entering and leaving by the side branches. We therefore give them separate labels I_1, I_2, I_3, I_4, and the total of the potential differences becomes:

$$I_1R_1+I_2R_2+I_3R_3+I_4R_4$$

For simplicity the symbol Σ is used to indicate summation over the elements 1, 2, 3 and 4; then the balance of the e.m.f. V against the total potential difference is written:

$$V = \Sigma IR \qquad\qquad\qquad (3.4)$$

This is the best known form of Kirchhoff's second law. In order to keep the signs correct here, we follow the closed path in either direction, counting a current as positive when it moves with us and an e.m.f. as positive when its source is traversed from negative to positive terminal. Using this convention a positive e.m.f. causes a positive current.

If the circuit of figure 3 is completed by inserting all the resistances and sources of e.m.f. attached to the side arms, then it should be possible to obtain expressions for the currents in terms of the e.m.f.s by applying Kirchhoff's second law to every closed path, and the first law to every junction. In practice it is often better to apply the first law at the beginning by labelling the currents in a way which must satisfy it: for example, the current leaving the junction of elements 2 and 3 by the side arm may be labelled (I_2-I_3).

Another practical hint for solving problems is that one of the closed paths in a complicated circuit may be made up by adding the others. Therefore the equation obtained by applying Kirchhoff's second law to the last closed path adds nothing to the information contained in the equations obtained from the other paths. The information we need is found by applying the law to all but one of the closed paths.

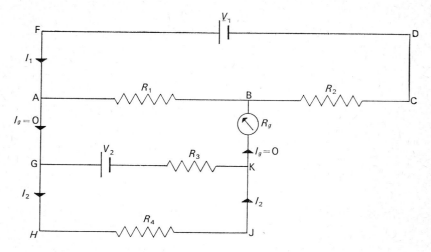

Figure 4.   Potentiometer circuit.

For example, in dealing with the potentiometer circuit of figure 4, we use three closed paths:

ABCDFA, with current I_1 flowing round it.
AGKBA,   ,,   ,, I_g ,,   ,,   ,,
GHJKG,   ,,   ,, I_2 ,,   ,,   ,,

These are the only currents flowing in the wires which form part of only one loop. However, R_1 forms part of two loops and carries a total current $I_1 - I_g$; similarly R_3 carries a current $I_2 - I_g$. When the galvanometer current I_g is zero there are only two unknown currents, I_1 and I_2.

Kirchhoff's second law applied to the path ABCDFA gives:

$$V_1 = I_1(R_1 + R_2) \tag{3.5}$$

Applied to the path GHJKG, it gives:

$$V_2 = I_2(R_4 + R_3) \tag{3.6}$$

Applied to the path GABKG, it gives:

$$V_2 = I_2R_3 + 0R_g + I_1R_1 \tag{3.7}$$

These three equations may be solved simultaneously to give I_1 and I_2 and one of the resistances R_2, R_3 or R_4, in terms of V_1, V_2 and the other resistances. Nothing is gained by applying the law to the big path AGHJKBCDFA, which just gives the sum of equations (3.5), (3.7) and (3.6) reversed.

3·7·3 Application to real cells

In practice cells are not the perfect sources of e.m.f. represented by the conventional symbol, as the e.m.f. across their terminals falls with increase in the current drawn from them. This behaviour is described by treating a real cell as a perfect source of e.m.f. in series with a resistance which is called the internal resistance of the cell. Thus in the circuit of figure 4, V_2 and R_3 might be the e.m.f. and internal resistance which we use together to represent the properties of a real cell. The circuit could be used as follows for the measurement of R_3 without knowledge of V_2:

With R_4 disconnected, a second balance point ($I_g = 0$) is obtained. In this case I_2 becomes zero, I_1 remains unchanged, R_1 becomes $R_1 + \delta$, R_2 becomes $R_2 - \delta$ and instead of equation (3.7) we have:

$$V_2 = I_1(R_1 + \delta) \tag{3.8}$$

Substituting for I_1 and I_2 in equation (3.7) by means of equations (3.6) and (3.8), we get:

$$V_2 = \frac{R_3}{R_4 + R_3} V_2 + \frac{R_1}{R_1 + \delta} V_2,$$

whence: $\qquad \dfrac{R_3}{R_4 + R_3} + \dfrac{R_1}{R_1 + \delta} = 1,$

which gives: $\qquad R_3 = \dfrac{R_4}{R_1} \delta.$

Chapter 4
The nature of the magnetic field

4·1 Perspective

In this chapter we consider a frequently ignored link in the logical structure of electromagnetic theory. The reader who finds it frightening, difficult, or just unconvincing, need not give up. The chapters which follow give a self-contained and fairly orthodox account of the magnetic effects of electric currents, starting from the Biot-Savart law and the expression for the force on a current element. This chapter serves to bind the subject into a whole by showing that these two laws are themselves logically linked (although not necessarily as consequences – see 'Preface') to the laws of electrostatics: magnetic effects are described as relativistic corrections to the electrostatic forces between moving charges.

Current electricity was understood and used as a self-contained subject before Einstein's work on relativity. Historically speaking, Einstein was led toward the special theory of relativity by a 'conviction that the electromotive force acting on a body in motion in a magnetic field was nothing else but an electric field' (from a letter quoted by R. S. Shankland, *Am. J. Phy.*, vol. 35, 1964, p. 32). But we, having grown accustomed to relativistic mechanics, have the possibility of using them as a tool in the construction of a complete electromagnetic theory. The initiative for developing and spreading the idea of magnetism as a relativistic correction to electrostatics is shared by many people, but special mention must be made of the work of W. G. V. Rosser (*Contemporary Physics*, vol. 1, 1959, p. 134; vol. 1, 1960, p. 453; vol. 3, 1961, p. 28).

4·2 Relativity of mechanical forces

4·2·1 *Time-dilatation*

Every day, around the big particle accelerators which are used for modern research in subnuclear physics, people carry out experiments on pi-mesons which have lived for times longer than 10^{-7} seconds. This is remarkable because pi-mesons are unstable particles, decaying with a mean life of $2 \cdot 5 \times 10^{-8}$ seconds when they are at rest. How is it that only a few percent of them decay during a journey of tens or hundreds of metres made at a velocity of the order of 3×10^8 metres per second?

The answer to this question is contained in the term 'time-dilatation'. Einstein's theory of Special Relativity, working from the experimental fact that the velocity of light appears the same to all observers, shows that the same interval of time may appear different to two observers. Especially let us consider a particle moving with velocity v relative to an observer O; a time-interval in the life-history of the particle, which appears to have a value t when measured in the rest-frame of the particle, will appear to the observer O to have a larger value γt, where:

$$\gamma = \frac{1}{\sqrt{\left(1 - \frac{v^2}{c^2}\right)}}$$

In the case of the pi-mesons the half-life measured in the laboratory, for decay in a beam of velocity v, is greater by a factor:

$$\gamma = \frac{1}{\sqrt{\left(1 - \frac{v^2}{c^2}\right)}}$$

than the half-life which would be measured by an observer moving with the beam. However, the latter observation is not easily made, so it is replaced by an equivalent measurement of the half-life of identical particles which have been brought to rest in the laboratory. It is only the relative velocity of observer and object that matters.

4·2·2 Transformation of force

Measurements of mass and length are also modified by relative velocity; these modifications are discussed in books on relativity (e.g. W. H. McCrea, *Relativity Physics*, Methuen, 1934). For our purposes it is necessary only to refer to the fact that observers in uniform relative motion agree about the magnitude of a momentum at right-angles to their relative motion.

Suppose that some outside force causes a moving particle (e.g. one of the pi-mesons in the beam mentioned above) to receive a small amount of momentum δp at right-angles to its velocity. From the Newtonian definition of force as rate of change of momentum, an observer moving with the particle who sees momentum δp acquired in a time δt_0, will say the force acting was $F_0 = \dfrac{\delta p}{\delta t_0}$. An observer in the laboratory, however, will attribute the same magnitude δp to the momentum but will think it was acquired over a longer time, namely:

$$\delta t = \gamma \delta t_0 = \frac{\delta t_0}{\sqrt{\left(1 - \frac{v^2}{c^2}\right)}}$$

and will therefore say that the force acting on the particle was:

$$F = \frac{\delta p}{\delta t} = \frac{\delta p}{\gamma \delta t_0} = \frac{1}{\gamma} F_0 \tag{4.1}$$

We thus see that an observer moving with the particle (i.e. in the 'rest-frame' of the particle) is the one who attributes the smallest value to a time-interval, and the largest value to a sideways force experienced by the particle.

4·3 Forces between electric charges in relative motion

In Chapter 1 we considered the forces between stationary electric charges, and in the next chapter the charges were allowed to move and the changes of potential energy when they did so were calculated. These calculations were made on the assumption that the velocity of the test-charge did not affect the force on it. In order to ensure the validity of this assumption we could have restricted our consideration to test-charges moving infinitely slowly.

However, in Chapter 3 we examined processes of conduction, in which the drift velocity of the electrons is a real velocity determined by the physical conditions. This drift velocity is usually small and is superimposed on normally greater, but still small, thermal velocities. Therefore, as a first approximation, we are able to use the arguments which apply to infinitely small velocities. In order to obtain a more rigorous description of the forces on electrons which are moving with significant velocities, we may invoke the language of section 4·2.

When two charges q and Q are at rest, the force on q is given by equation (2.3) as:

$$\mathbf{F} = q\mathbf{E} \tag{4.2}$$

where $\mathbf{E}$ is the electric field due to Q at the position of q, given by equation (2.4) as:

$$\mathbf{E} = \frac{1}{4\pi\varepsilon_0} \frac{Q}{r^2} \frac{\mathbf{r}}{r} \tag{2.4}$$

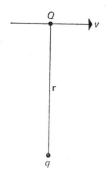

Figure 5.   Source-charge moving, test-charge at rest.

Now suppose that q remains at rest (see figure 5) while Q and the observer are moving with velocity v at right-angles to the line joining Q and q. If measurements are made by this observer, the electric field produced by Q at the position of q is still $\mathbf{E}$ and the force on q is still $q\mathbf{E}$, independent of v.* But to an observer at rest in the laboratory, which is the rest-frame of q, the force on q will appear to be larger by a factor γ: this follows from equation (4.1). If the force on q, as measured by the stationary observer, is written:

$$\mathbf{F_0} = \gamma\,\mathbf{F} = \gamma\,q\,\mathbf{E} = q\,\mathbf{E_0}, \tag{4.3}$$

we see that this observer will describe the situation by saying that there is, in the neighbourhood of q, an electric field

$$\mathbf{E_0} = \gamma\,\mathbf{E} \tag{4.4}$$

Thus different observers, each defining electric field as force per unit charge on a test-charge at rest with respect to himself, will attribute to the electric field at a point, values which differ in the same way as do the values attributed to a force.

4·4   Force between two moving charges

4·4·1   *Total force*

If we have the charges q and Q both moving with velocity v in the same direction, at right-angles to the line joining them (see figure 6), the force on q (measured by an

Figure 6.   Both charges moving with same velocity.

observer moving with the charges) will be $q\mathbf{E}$. But this observer is now in the rest-frame of q; so according to equation (4.1), an observer at rest in the laboratory will attribute to the force on q a smaller value, namely:

---

* For this to hold, electric charge must be relativistically invariant, appearing the same to all observers. Experiments provide plenty of indirect experimental evidence for this invariance but perhaps the most direct evidence for it is the fact that the total charge of a system does not depend upon the velocities of its charged components; atoms are electrically neutral in spite of the velocities of their electrons.

$$\mathbf{F'} = \frac{1}{\gamma} q \, \mathbf{E} \qquad\qquad (4.5)$$

4·4·2 Comparison to the force on a stationary charge

The force $\mathbf{F}_0$ on a similar stationary charge at the same place and time is given by equation (4.3), so if we want to describe the force on the moving q as a force $\mathbf{F}_0$ which is independent of its movement, plus an extra force $\mathbf{F}_m$ associated with its movement, $\mathbf{F}_m$ must be obtained by subtracting equation (4.3) from equation (4.5), as follows:

$$\mathbf{F}_m = \mathbf{F'} - \mathbf{F}_0 = q\,\mathbf{E}\left(\frac{1}{\gamma} - \gamma\right)$$

$$= q\,\mathbf{E}\left\{\left(1 - \frac{v^2}{c^2}\right)^{\frac{1}{2}} - \left(1 - \frac{v^2}{c^2}\right)^{-\frac{1}{2}}\right\}$$

For sufficiently small values of $\frac{v}{c}$, we may expand the terms on the right-hand side by the binomial theorem, ignoring all terms higher than the first power of $\frac{v^2}{c^2}$. This gives:

$$\mathbf{F}_m = -q\,\mathbf{E}\frac{v^2}{c^2} \qquad\qquad (4.6)$$

We thus see that the total force on the moving charge q may be expressed as the resultant of two forces: $\mathbf{F}_0$ which is independent of its velocity, and $\mathbf{F}_m$ which gives the effect of the velocity; the minus sign indicates that the extra force $\mathbf{F}_m$ is in the opposite sense to the electrostatic force $\mathbf{F}_0$. Since material objects cannot move with velocities larger than that of light, the factor $\frac{v^2}{c^2}$ is always less than unity. This means that when Q and q are of the same sign, making $\mathbf{F}_0$ a repulsion, $\mathbf{F}_m$ is an attraction which can reduce the magnitude of the net repulsive force, but can never lead to a net attraction, however great the velocity.*

4·4·3 The magnetic force

The suffix m has been given to the extra force, $\mathbf{F}_m$, because the latter is usually attributed to magnetic effects. The moving source-charge Q is said to produce a magnetic field with flux density B. The characteristic of this field is that a test-charge q moving through it with velocity v experiences a force

* In the 'pinch effect', an electric discharge in a gas is observed to contract laterally, because the normal electrostatic repulsion between electrons is cancelled by the effects of positive ions in the gas: when movement reduces the total repulsive force on the electrons, the net force on them is an attraction equal to $\mathbf{F}_m$. The role of positive ions in determining the forces between currents in wires is similar to that in the pinch effect and is discussed in section 4·6.

$$F_m = qvB \qquad \textbf{(4.7)}$$

Fitting equation **(4.7)** to equation **(4.6)**, we see that the moving charge Q must produce a magnetic field with flux density

$$B = \frac{Ev}{c^2} \qquad \textbf{(4.8)}$$

Although equations **(4.7)** and **(4.8)** relate the magnitude of vectors, they are not written in vector notation because they do not yet relate the directions of the vectors; we shall see in section 4·6 that B must be considered as a vector perpendicular to E and to the velocity.

4·5 Definition of magnetic field

4·5·1 The case of unequal velocities

In the calculation of the magnetic force between two charges with the same velocity in the previous section, one factor v was included in the expression **(4.8)** for B and another in equation **(4.7)** which gave the force on a charge moving through a region of given magnetic flux density B. This device not only describes the special case of equal velocities, but also gives the correct result (no magnetic force) when either velocity is zero. A little algebra shows that it also correctly covers the general case of unequal velocities, if v in equation **(4.7)** is taken to be the velocity of q, but u is put as the velocity of Q in equation **(4.8)**, which becomes:

$$B = \frac{1}{c^2} Eu \qquad \textbf{(4.9)}$$

4·5·2 The case of unequal velocities in different directions

The still more general case, in which the charges move in different directions, requires careful treatment by relativistic vector methods. The calculations involved are rather laborious when the velocities have components parallel to E, but a calculation using three-dimensional vectors with relativistic rules for adding them, is given by Rosser (*An Introduction to the Theory of Relativity*, 1964, Butterworth, pp. 285–90). An alternative calculation, using four-dimensional vectors, is given in Appendix C. These calculations both lead to the normal solution, which is to consider B as a vector perpendicular to E and u; it is obtained by writing the right-hand side of equation **(4.9)** as a vector product (see Appendix B):

$$\mathbf{B} = \frac{1}{c^2}\, \mathbf{u} \times \mathbf{E} \qquad \textbf{(4.10)}$$

In Chapter 9, we shall define an electric flux density D, equal to $\varepsilon_0 E$ in free space. We may therefore replace E in equation **(4.10)** by $\dfrac{\mathbf{D}}{\varepsilon_0}$, obtaining:

$$B = \frac{1}{\varepsilon_0 c^2} u \times D = \mu_0 u \times D \tag{4.11}$$

This introduces a new constant μ_0, related to ε_0 by:

$$\mu_0 = \frac{1}{\varepsilon_0 c^2} \tag{4.12}$$

In fact μ_0 is given the value

$$\mu_0 = 4\pi \times 10^{-7} \text{ newton ampere}^{-2} \text{ or henry m.}^{-1}$$

for reasons discussed on page 54, and the value of ε_0 follows from this. We shall reserve the name magnetic field for the vector $H = \dfrac{B}{\mu_0}$, but it should be noted that some authors prefer to call B the magnetic field, while others simply use the names H-field or B-field. In terms of H, equation (4.11) becomes:

$$H = u \times D \tag{4.13}$$

This is a conventional expression for the magnetic field due to a moving charge, in terms of the electric field which it causes in its own rest-frame. If we wish to express H in terms of the magnitude and position of the moving charge, we may substitute equation (2.4) into equation (4.13) to get:

$$H = \frac{1}{4\pi} \frac{Q}{r^3} u \times r \tag{4.14}$$

4·5·3   The Lorentz force

Now that we have defined the magnetic field, in direction as well as magnitude, we must re-write equation (4.7) in vector form: to obtain F_m along the same line

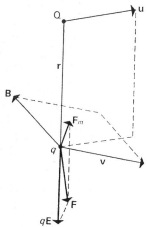

Figure 7.   The force on a moving test-charge q due to a moving source-charge Q.
B is perpendicular to the plane of r and u;
F_m is perpendicular to the plane of B and v;
F is the resultant of qE and F_m.

as **E**, but in the opposite sense, vB must be replaced by a vector product $\mathbf{v} \times \mathbf{B}$, giving:

$$\mathbf{F}_m = q\,\mathbf{v} \times \mathbf{B} \tag{4.15}$$

This too is a well-known equation and the force which it describes is often called the Lorentz force. Many chapters on electromagnetism start by stating that the force on a charge q, moving with velocity v through a region where unspecified sources lead to the presence of an electric field **E** and a magnetic flux density **B** is:

$$\mathbf{F} = q\,(\mathbf{E} + \mathbf{v} \times \mathbf{B}) \tag{4.16}$$

The relative directions of all these vectors are illustrated in figure 7.

4·6 Forces between wires carrying currents

As we have seen in Chapter 3 a wire carrying an electric current must contain one fixed positive ion for each moving electron. These ions, as well as the electrons, must be considered in the calculation of the force between two wires in which currents are flowing.

Consider two stationary parallel wires with one positive ion and one electron in each wire. The ion and the electron in a given wire are at positions so close to each other that they may be considered to be at the same place. The ion in wire A and the ion in wire B lie on a line perpendicular to the length of the wires and both electrons are moving with velocity v parallel to the length of the wires. The forces on wire A, as measured by a stationary observer, are as follows:

The total force due to the positive ion in B is zero, because the measurements are being made by an observer in the rest-frame of the source-particle: hence by the argument of section 4·3, such an observer will see equal and opposite forces on equal and opposite charges at the same place, whether or not they are in motion.

On the other hand, the forces on the ion and the electron in A, due to the electron in B, do not cancel, as the force on the electron is $\mathbf{F}'$ of equation **(4.5)**, while the force on the ion is $\mathbf{F}_0$ of equation **(4.3)** (with sign changed to allow for the opposite charge). The total of these is a repulsion of magnitude $\mathbf{F}' - \mathbf{F}_0$, which has been written in equation **(4.6)** as an attraction of magnitude:

$$\mathbf{F}_0 - \mathbf{F}' = \mathbf{F}_m = q\mathbf{E}\,\frac{v^2}{c^2}$$

We thus see that the magnetic force $\mathbf{F}_m$, which was introduced in section 4·5, is a good basis for discussing the net forces between currents flowing in conductors. The direct, physical significance of this statement is that the stationary positive ions serve to cancel that part of the force on the electrons which might be attributed to an electrostatic field, leaving only the extra term associated with the velocity.

Whether we call the above an explanation, or a description, of the nature of magnetic forces, it is a direct demonstration of the fact that magnetic forces are smaller by a factor $\dfrac{v^2}{c^2}$ (about 10^{-28} for the example of section 3.1) than the electrostatic forces which would exist if the electrons (or the ions) were present alone. In the following section we shall see that magnetic forces are of practical importance simply because the numbers of moving electrons present in current-carrying wires are larger than the excess of electrons over ions commonly observed on electrically-charged bodies, by factors of order $\dfrac{c}{v}$.

4·7   The magnetic field due to a current element

Let us consider a wire containing n mobile electrons per unit length, all moving with the same drift velocity u. Each electron carries a charge $-e$ and produces at a distant point P a magnetic field given by equation (4.14) as:

$$\mathbf{H} = \frac{1}{4\pi} \frac{-e}{r^3} \mathbf{u} \times \mathbf{r}$$

where $\mathbf{r}$ specifies the position P with respect to the electron (see figure 8).

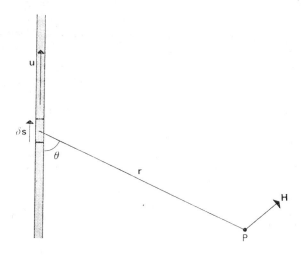

Figure 8.   Magnetic field due to a current element.

An element of the wire of length δs will contain a number of electrons $n\,\delta s$ and will therefore produce at P a total magnetic field:

$$\mathbf{H} = \frac{1}{4\pi} \frac{-e}{r^3} n\delta s \, \mathbf{u} \times \mathbf{r} \tag{4.17}$$

As we have a length δs measured along the direction of the wire multiplied by a velocity vector $\mathbf{u}$ parallel to the direction of the wire, it is possible, without changing the significance of equation (4.17), to make the length a vector $\delta\mathbf{s}$ and multiply it by u, the magnitude of the velocity vector; this gives:

$$\mathbf{H} = \frac{1}{4\pi} \frac{-e}{r^3} n \, u \, \delta\mathbf{s} \times \mathbf{r} \tag{4.18}$$

It can now be seen that $(-enu)$ is the total charge per second passing a given point in the wire; this is the quantity to which we usually give the name electric current and the symbol I. If e is measured in coulombs, I is the current in coulombs per second, which are called amperes. The magnetic field at P may therefore be written as:

$$\mathbf{H} = \frac{1}{4\pi} \frac{I}{r^3} \delta\mathbf{s} \times \mathbf{r} \tag{4.19}$$

The notation of figure 8 can be used to write the vector product $\delta\mathbf{s} \times \mathbf{r}$ as $r\delta s \sin\theta$, which gives the magnetic field at P as:

$$H = \frac{1}{4\pi} \frac{I\delta s \sin\theta}{r^2} \tag{4.20}$$

This is the Biot-Savart law commonly used (in this, as in other books) as a starting-point for calculating the magnetic field due to currents flowing in real apparatus – coils, solenoids and so on. We shall use it as the basis for Chapter 5.

The right-hand sides of equations (4.19) and (4.20) have the dimensions of current per unit length. Thus the equations serve to define a unit of magnetic field which is called the ampere per metre. In words, we may say that a magnetic field of one ampere per metre would be produced if a current of 4π amperes flowed in an arc of length one metre, of a wire in the form of a circle of radius one metre. Anticipating equation (5.6), this means that a current of 2 amperes flowing round the circumference of a single circular turn of radius 1 metre produces at the centre a magnetic field of 1 ampere per metre.

To those who are used to thinking of magnetic fields measured in oersteds, we point out that:

$$1 \text{ ampere per metre} = 4\pi \times 10^{-3} \text{ oersted.}$$

It is thus a rather small unit; the horizontal component of the earth's magnetic field in Britain, which is about 0·18 oersted, is about 14 amperes per metre. One oersted is close to 80 amperes per metre (see Appendix A).

4·8 **Force on a current element in a magnetic field**

In section 4·7, we made the transition from a single charge to a current element for the source of the magnetic field. In this section the corresponding transition is made for the object experiencing the force.

We start from equation (4.15), which gives the force on a single electron moving with velocity $\mathbf{v}$ in a magnetic field of flux density $\mathbf{B}$ as:

$$\mathbf{F} = (-e)\mathbf{v} \times \mathbf{B}$$

Then as in section 4·7, we consider a wire containing n such electrons per unit length, so that an element of length δs contains $n\delta s$ electrons and therefore experiences a total force:

$$\mathbf{F} = (-e)n\delta s \ \mathbf{v} \times \mathbf{B}$$

which we re-write by the device of section 4·7 as:

$$\mathbf{F} = (-e)n \ v \ \delta \mathbf{s} \times \mathbf{B}$$

$$= I\delta \mathbf{s} \times \mathbf{B} \tag{4.21}$$

This is a standard expression for the force on an element of wire carrying a current I. It shows that the force per unit length of wire is perpendicular to the wire and to the magnetic field, its magnitude being given by the product of I with the component of $\mathbf{B}$ in the direction perpendicular to the wire. This discussion will be continued in Chapter 6.

Chapter 5
Calculations of magnetic field

5·1 **Magnetic field of a long straight wire**

5·1·1 Equation **(4.20)** gives an expression for the magnetic field of a short element of a wire carrying a current. Such an element cannot, of course, exist in isolation, for the current must somehow enter and leave it. A current flowing in a real electric circuit will give, at any point, a total magnetic field which may be obtained by summing the contributions from all the elements of wire which make up the circuit. In the limit of infinitely short elements the summation becomes integration. We shall start the present chapter by showing how this integration is carried out for several simple circuits or parts of circuits.

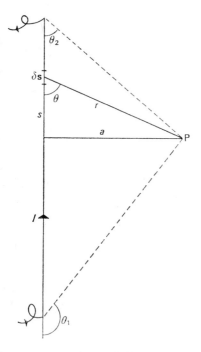

Figure 9.   Magnetic field due to a straight wire

5·1·2   Before proceeding to complete circuits we consider the case of a long straight wire fed with current through connections which are so far away that they can be ignored. In the notation of figure 9, equation (4.20) gives the magnetic field at P due to the element δs as:

$$\delta H = \frac{I \delta s \sin \theta}{4 \pi r^2} \tag{5.1}$$

in a direction perpendicular to the plane of the figure. The magnetic field of the whole wire at P is therefore:

$$H = \int \frac{I ds \sin \theta}{4 \pi r^2} \tag{5.2}$$

again in a direction perpendicular to the plane of the figure. To evaluate this integral, we must express two of the variables s, r and θ in terms of the other, so that we have an integral containing only one variable. If θ is chosen as our one variable, s and r may be eliminated by means of the relations:

$$r = a \operatorname{cosec} \theta$$

and:

$$s = a \cot \theta$$

whence:

$$ds = \frac{ds}{d\theta} d\theta = -a \operatorname{cosec}^2 \theta \, d\theta.$$

With these substitutions, the integral becomes:

$$H = \int \frac{I \sin \theta (-a \operatorname{cosec}^2 \theta) d\theta}{4 \pi a^2 \operatorname{cosec}^2 \theta}$$

$$= \int \frac{-I}{4 \pi a} \sin \theta \, d\theta$$

Here a is the perpendicular distance of P from the wire, which is not a variable: therefore we may put $\dfrac{I}{4 \pi a}$ outside the integral sign. We may also insert the limits of integration θ_1 and θ_2, representing the values of θ at the ends of the wire, to obtain:

$$H = \frac{I}{4 \pi a} \int_{\theta_1}^{\theta_2} (-\sin \theta) \, d\theta$$

$$= \frac{I}{4 \pi a} \int_{\theta_1}^{\theta_2} d(\cos \theta)$$

$$= \frac{I}{4 \pi a} (\cos \theta_2 - \cos \theta_1) \tag{5.3}$$

37   Calculations of magnetic field

If the wire is infinitely long, $\theta_2 = 0$ and $\theta_1 = 180°$, so the difference of the cosine terms is 2, and we get:

$$H = \frac{I}{2\pi a} \tag{5.4}$$

This magnetic field is at right-angles to the plane containing P and the wire; it follows that the lines of magnetic field are circles around the wire, the field around each circle being inversely proportional to its radius.

The magnetic field of a rectangular coil, or of a complete circuit made up of straight wires, may be obtained by using equation (5.3) for each wire.

5·2 Magnetic field of a circular coil

5·2·1 When we calculate the magnetic field of a wire bent into the form of a coil, we neglect the magnetic effects of the connections through which the current enters and leaves the coil. If we had stressed this point in the case of the long straight wire, we might have been troubled by the fact that the connections, though very far away, were also very long and not necessarily negligible in practice. However, in the case of a coil, the current may be made to enter and leave at points very close together. Therefore the connections, which carry equal and opposite currents, may be laid parallel, or twisted together, so close that their total magnetic field is equal to that of a single conductor carrying zero current, i.e. is zero in both theory and practice. Thus an isolated current-carrying coil, while it contains an element of abstraction, is something that can be realized in practice to a very good approximation. This is in marked contrast to the isolated current elements, which are essentially unrealizable, having a usefulness which is logical rather than directly experimental. Between these two extremes there is the isolated long straight wire, which is an ideal realizable in practice to a rough approximation.

5·2·2 We now use equation (4.20) as a basis for calculating the magnetic field at a

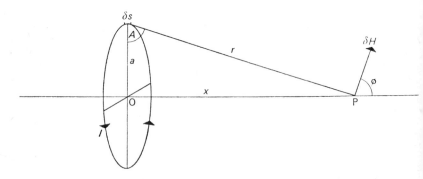

Figure 10. Magnetic field due to an element of a circular coil.

point P, on the axis of a circular coil. By the notation of figure 10, an element of the coil gives a magnetic field at P of:

$$\delta H = \frac{I\delta s}{4\pi r^2}$$

There is no need to include the factor $\sin\theta$ since all the elements of the coil are perpendicular to the lines AP joining them to P. We must, however, consider the direction of the field δH; it is in the plane containing AP and OP, in the direction defined by the angle ϕ. There is therefore a component $\delta H \sin\phi$ perpendicular to the axis OP and a component $\delta H \cos\phi$ along the axis.

When the effects of all the elements of the coil are summed, the components perpendicular to the axis cancel; the components along the axis add, however, to give a net field along the axis of magnitude:

$$H = \Sigma \frac{I\delta s}{4\pi r^2} \cos\phi$$

$$= \Sigma \frac{I\delta s}{4\pi r^2} \frac{a}{r}$$

$$= \frac{Ia}{4\pi r^3} \Sigma \delta s$$

If the coil consists of n complete turns, making $\Sigma \delta s = 2\pi n a$, this gives:

$$H = \frac{nIa^2}{2r^3} \qquad (5.5)$$

Equation (5.5) will be used in section 5·3 to give the field inside a solenoid, but for the time being we shall simply point out two special cases:

(i) At the centre of the coil, where P is the same as O, $r = a$ and (5.5) becomes the well-known expression:

$$H = \frac{nI}{2a} \qquad (5.6)$$

(ii) At a distant point on the axis r^3 may be replaced by x^3 and the field is

$$H = \frac{nIa^2}{2x^3} = \frac{IA}{2\pi x^3}, \qquad (5.7)$$

where A is the area of the coil, each turn being counted separately. It may be shown that equation (5.7) holds for any shape of coil, provided the whole of the area A is close to the axis (close in comparison with the distance x).

3   **Magnetic field inside a solenoid**

A solenoid is a spiral of wire tightly wound on a cylindrical former so that the turns are very nearly perfect circles with planes perpendicular to the axis of the cylinder. When we say that they are tightly wound, we mean that the axial spacing of the turns is much less than the radius of the cylindrical former.

Let us assume that the solenoid is wound uniformly, with m turns per unit length; at a point P on the axis, the field due to a single turn is given by equation

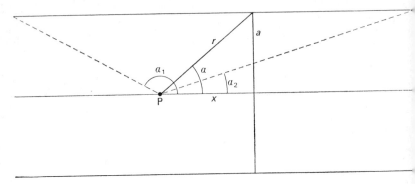

Figure 11. Calculation of magnetic field inside a solenoid.

(5.5), in the notation of figure 11, as:

$$H = \frac{Ia^2}{2r^3}$$

Since this field is along the axis, the field due to the whole solenoid may be calculated by adding the contribution of all the turns, each with the appropriate value of r. The total field is therefore:

$$H = \frac{Ia^2}{2} \Sigma \left(\frac{1}{r^3} \right)$$

$\Sigma \left(\frac{1}{r^3} \right)$ is troublesome to obtain by direct summation; but if the number of turns is large a very good approximation can be obtained by integrating over the length of the solenoid, just as if the turns were distributed continuously at a density of m turns per unit length. The resulting expression for the field at P is:

$$H = \frac{Ia^2}{2} \int r^{-3} \, m dx$$

By expressing both r and x in terms of the constant a and the variable angle α in a manner similar to that used for the long straight wire, we convert this to

$$H = \frac{Ia^2}{2} \int \frac{\sin^3 \alpha}{a^3} \, m d(a \cot \alpha)$$

$$= \frac{mI}{2} \int \sin^{\cdot} \alpha (-\operatorname{cosec}^2 \alpha) d\alpha$$

$$= \frac{mI}{2} \int d(\cos \alpha)$$

$$= \frac{mI}{2} (\cos \alpha_2 - \cos \alpha_1) \tag{5.8}$$

It follows that when the solenoid is infinitely long in comparison with its radius, making $\cos \alpha_2 = 1$ and $\cos \alpha_1 = -1$, the field at P is:

$$H = mI \tag{5.9}$$

In practice solenoids are not infinitely long and so for each point on the axis of a real solenoid this simple expression must be multiplied by a correcting factor $\frac{1}{2}(\cos \alpha_2 - \cos \alpha_1)$. This gives a result which is still valid and easily calculated for points near, or even outside, the ends of the solenoid; it shows that the field at one end of a moderately long solenoid is half that in the middle.

5·4   Line integral of magnetic field

4·1   In section 2·3, we considered the line integral of electric field $\int E \cdot ds$. This was useful because, for a path between two points, it gave the work done in carrying a unit electric charge between the two points. For a closed path, it was necessarily zero.

We are now going to consider the line integral of magnetic field $\int H \cdot ds$ which is mathematically rather similar, but useful for quite different reasons. In the first place, when it is taken between two points it does not represent the work done in moving any known type of object, since free magnetic charges (or poles) do not exist; secondly, when taken round a closed path it has a value proportional to the magnitude of any current which may be flowing through the path. This value is not in general zero and we shall now proceed to evaluate the constant of proportionality.

4·2   As the simplest example, let us take a circular path round a long straight wire. Such a path is illustrated in figure 12, from which it may be seen that H is along the path all the way round and of uniform magnitude $\dfrac{I}{2\pi a}$. It follows that the line integral of H is simply the magnitude of H multiplied by the circumference of the circular path, which is $2\pi a$; the factors $2\pi a$ cancel and the result is simply I. Therefore if the symbol $\oint$ is used to represent line integral around a closed path, the conclusion may be written:

$$\oint H \cdot ds = I \tag{5.10}$$

4·3   We shall now show that equation (5.10) holds for any closed path round a long straight wire: one such path with a typical element δs is illustrated in figure 12.

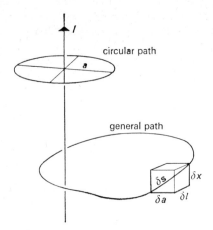

Figure 12.   Line integral of magnetic field round a closed path.

Instead of resolving the element **δs** of the path into components along fixed x-, y- and z-axes, we resolve it into three mutually perpendicular components: a radial component δa, an axial component δx (parallel to the wire) and a tangential component δl (parallel to **H**). δa and δx contribute nothing to the value of **H.δs**, which may therefore be written as:

$$\mathbf{H.\delta s} = H\delta l$$

$$= Ha\delta\theta$$

$$= \frac{I}{2\pi a}\, a\delta\theta$$

$$= \frac{I}{2\pi}\, \delta\theta$$

To integrate, or sum, this over the whole path is very easy, as every element contributes $\dfrac{I}{2\pi}$ times the angle $\delta\theta$ which its tangential component subtends at the wire. For the whole path, the total of all such elements of angle $\delta\theta$ is 2π, even in the limit of infinitesimal elements. Therefore the line integral is:

$$\oint \mathbf{H.ds} = \frac{I}{2\pi}\times 2\pi = I,$$

which is equation (5.10) again.

If a path is chosen which goes round a number of wires, or round a region carrying a distributed current, equation (5.10) holds with I representing the total current through the path.

Equation (5.10) can now be shown to be even more general in its application, if the restriction to paths round long straight wires is removed (for the purposes of the above arguments, 'long straight' means 'straight for a length which is great in comparison with the dimensions of the path'). Even if the wire is curved this condition can be met by choosing a very small path round it. Therefore if the wire is bent very sharply, it may be necessary to start with a large number of adjacent small paths inside it. These small paths are arranged to make up a complete net, so that the sum of the line integrals round them is the line integral round the single path consisting of their outer sections; the inner sections contribute nothing to the sum, because each contributes positively to $\oint$ **H.ds** for one small path and negatively to $\oint$ **H.ds** for the path on the other side of it. Since the current I flowing through the wire is the sum of the currents flowing through all the little paths, and as the small paths can be made small enough to ensure that the current through each flows straight for a distance great in comparison with the dimensions of the path, it follows that $\oint$ **H.ds** for a path just outside the wire is equal to I. This path can now be converted to a larger one, of any required shape or size, by adding more small paths to the outside of it. Again, sections common to two paths cancel and the sum of the line integrals is equal to that for the path made up of the outside boundary of the net. But since the extra paths are outside the wire, and no current flows through them, they add nothing to the total value of $\oint$ **H.ds**.

We thus reach the conclusion that the line integral of magnetic field, round *any* closed path, is numerically equal to the current flowing through the path. If the current I flows through a path n times, for example because the path encloses n turns of a coil, then the effective current flowing through the path is nI, and we must put:

$$\oint \mathbf{H.ds} = nI \tag{5.11}$$

We have reached the above conclusions through an argument which involves the abstract idea of making up a closed path out of a network of smaller paths, the details of which do not matter. It is possible to reach the same conclusion by other arguments, many of which involve the additional abstraction of a magnetic shell which has boundary coincident with the wires of the circuit, and external magnetic field equivalent to that of the circuit.

It is characteristic of electromagnetic theory that facts are often related by more than one type of argument; all may be valid and equally 'fundamental', but individual opinion differs as to which is the most satisfactory. Equally it is characteristic that we often have freedom to choose which of the two ends of an argument we consider the more basic and thus more suitable for use as a starting-point from which the other is derived. The subject matter of the present section is often treated by taking equation (5.10) as starting-point, but equally often the equivalent magnetic shell is used as the basis for calculating the magnetic fields of current-carrying wires, equation (4.20) being derived from it.

5·5 Maxwell's equation for curl H

5·5·1 In the preceding section we have discussed the numerical equality of the line integral of magnetic field around a path and the current flowing through the path. Our discussion included the case where the current was confined within a wire of dimensions less than that of the path, and the case of a small path inside a wire, within which flowed only a part of the total current. If the current I were distributed uniformly over the cross-sectional area S of the wire, we could say that there was a current density:

$$i = \frac{I}{S}$$

Then through a path of area δS there would be a current $j\delta S$, which would be numerically equal $\oint \mathbf{H}.\mathbf{ds}$ round the path, so we could write:

$$\frac{\oint \mathbf{H}.\mathbf{ds}}{\delta S} = j \tag{5.12}$$

In words, this means that the current density is equal to the quantity obtained by taking the line integral of magnetic field round a path enclosing an area δS, and dividing by δS. This quantity must be a property of the magnetic field in the region, independent of the size of the path. It may be defined for regions of decreasing size, by taking smaller and smaller paths, until in the limit we have reached a region so small that it can be treated as a point.

5·5·2 Though we did not emphasize the assumption, in the preceding paragraph our path of area δS was taken in the plane perpendicular to the direction of current flow. It is now possible to dispense with this restriction by considering the case where the current, of density j, flows in a direction making an angle ϕ with the normal to the plane of our small path, which has area δS. In this case the current flowing through the path is $j\delta S \cos \phi$ and equation **(5.10)** requires that:

$$\oint \mathbf{H}.\mathbf{ds} = j\delta S \cos \phi$$

i.e.
$$\frac{\oint \mathbf{H}.\mathbf{ds}}{\delta S} = j \cos \phi \tag{5.13}$$

However the current density, which has magnitude j, has a direction also and may be represented by a vector $\mathbf{j}$. $j \cos \phi$ is simply the component of this vector along the normal to the area δS, and equation **(5.13)** requires that the component of $\mathbf{j}$ in *any* direction shall be equal to the value of $\dfrac{\oint \mathbf{H}.\mathbf{ds}}{\delta S}$ taken for a path of area δS in the plane normal to that direction.

From this we may deduce two facts:

(i) The quantities $\dfrac{\oint \mathbf{H}.\mathbf{ds}}{\delta S}$ are themselves the components of a vector, each in

the direction normal to the plane of the path used. If we consider many orienta-
tions of the path, we shall find one which gives a maximum value for $\dfrac{\oint \mathbf{H} \cdot \mathbf{ds}}{\delta S}$. The
magnitude of the vector is this maximum value of $\dfrac{\oint \mathbf{H} \cdot \mathbf{ds}}{\delta S}$ and its direction is
normal to the path which gives the maximum value. This vector is derived from
the vector $\mathbf{H}$ and is called the curl of $\mathbf{H}$, or rotation of $\mathbf{H}$, written curl $\mathbf{H}$ in
English-language texts and rot $\mathbf{H}$ in most other languages. (The English language
does, however, make one small concession to international usage in employing
the adjective irrotational to describe the fields of those vectors which have curl
equal to zero.)

It is shown in Appendix B that the curl of a vector may alternatively
be defined in terms of the differential coefficients of its components, and also that
the above argument may be generalised by using Stokes's theorem, which relates
the line integral of a vector to the surface integral of its curl.

(ii) Having defined curl $\mathbf{H}$, we may write equation (5.13) in vector form as:

$$\text{curl } \mathbf{H} = \mathbf{j} \tag{5.14}$$

This is one of the four relations known as Maxwell's equations. We shall meet
the other three in Chapters 7, 8 and 9. The four equations together are commonly
taken as a complete and concise description of the properties of electromagnetic
fields.

5·3   Equation (5.14) is in fact not quite complete. It tells us something about the
magnetic field in a region through which current is actually flowing. To make it
complete, a term for the magnetic field of moving source-charges which are
further away should be added. This term may be obtained directly from equation
(4.13):

$$\mathbf{H} = \mathbf{u} \times \mathbf{D} \tag{4.13}$$

This equation refers to a source-particle moving with velocity $\mathbf{u}$, and relates the
magnetic field $\mathbf{H}$ created by the particle, at any point, to its electric flux density
at that point.

The extra term needed to complete equation (5.14) is just the curl of $(\mathbf{u} \times \mathbf{D})$,
summed over all the moving source-particles which have not been included in
the current density $\mathbf{j}$. For a given source-particle, curl $(\mathbf{u} \times \mathbf{D})$ is a vector with
x-component equal to:

$$\frac{\partial}{\partial y}(u_x D_y - u_y D_x) - \frac{\partial}{\partial z}(u_z D_x - u_x D_z)$$

Here D_x, D_y and D_z are the components of electric flux density at a point x, y, z,
and are subject to differentiation by the operators $\dfrac{\partial}{\partial y}$ and $\dfrac{\partial}{\partial z}$. But u_x, u_y and u_z,
which are the components of the velocity of a source-particle, are not affected by

changing the coordinates x, y and z of a point of observation, and are therefore not subject to differentiation by the operators $\dfrac{\partial}{\partial y}$ and $\dfrac{\partial}{\partial z}$. The x-component of curl $(\mathbf{u} \times \mathbf{D})$ may thus be developed as:

$$\text{curl } (\mathbf{u} \times \mathbf{D})_x = u_x\left(\frac{\partial D_y}{\partial y} + \frac{\partial D_z}{\partial z}\right) - u_y\frac{\partial D_x}{\partial y} - u_z\frac{\partial D_x}{\partial z}$$

$$= u_x\left(\frac{\partial D_x}{\partial x} + \frac{\partial D_y}{\partial y} + \frac{\partial D_z}{\partial z}\right) + \left(-u_x\frac{\partial D_x}{\partial x} - u_y\frac{\partial D_x}{\partial y} - u_z\frac{\partial D_x}{\partial z}\right)$$

The first bracket in this expression is div $\mathbf{D}$ (see page 97). The second bracket is just the rate of change of the value D_x at the fixed point of observation, as the source particle goes past with velocity $\mathbf{u}$.

We may therefore write:

$$\text{curl } (\mathbf{u} \times \mathbf{D})_x = u_x \text{ div } \mathbf{D} + \frac{dD_x}{dt}$$

Since corresponding arguments apply to the y- and z-components, the total effect of a single source-particle is given by the vector equation:

$$\text{curl } (\mathbf{u} \times \mathbf{D}) = \mathbf{u} \text{ div } \mathbf{D} + \frac{d\mathbf{D}}{dt}$$

If we sum over all the source-charges, div $\mathbf{D}$ is the local charge-density (see page 98), and $\mathbf{u}$ div $\mathbf{D}$ is the current density $\mathbf{j}$, while $\dfrac{d\mathbf{D}}{dt}$ is the rate of change of total $\mathbf{D}$ due to source-particles not included in $\mathbf{j}$.

We have thus confirmed the correctness of the term $\mathbf{j}$ in equation (5.14), and shown that the complete form of Maxwell's equation for curl $\mathbf{H}$ is:

$$\text{curl } \mathbf{H} = \mathbf{j} + \frac{d\mathbf{D}}{dt} \tag{5.15}$$

The extra term $\dfrac{d\mathbf{D}}{dt}$ is sometimes given the inappropriate name of displacement current.

Its physical significance may be summarized as follows: if we measure the value of $\mathbf{D}$ at a single point and find it to be changing with time, then we assume that it is due to a moving source-particle. However, unless we measure the spatial variation of $\mathbf{D}$ as well, we cannot deduce the velocity $\mathbf{u}$ of the source-particle. But we can say $\mathbf{u}$ is such that the curl of $(\mathbf{u} \times \mathbf{D})$ is equal to the observed rate of change of $\mathbf{D}$. Also we can be sure that the moving source-particle is producing, at the point of observation, a magnetic field $\mathbf{H}$ with curl $\mathbf{H}$ equal to $\dfrac{d\mathbf{D}}{dt}$.

Chapter 6
Forces due to magnetic fields

6·1   **The unit of magnetic flux density**

We concluded Chapter 4 with an expression for the force on a current element in a magnetic field. If the element has length and direction represented by the vector δs, and carries a current I in a magnetic field which has flux density **B**, the force on it is:

$$\mathbf{F} = I\delta\mathbf{s} \times \mathbf{B} \tag{4.21}$$

This can be very simply extended to cover the case of a straight wire of length l, carrying a current I in a region of uniform magnetic flux density **B**: each element of the wire, of length represented by the small vector δs, experiences a force $I\delta\mathbf{s} \times \mathbf{B}$. The total force on a wire whose length is represented by the vector l is equal to the sum of the forces on all the short elements δs which make up the length l. Since the sum of the vector products of several vectors with **B** is equal to the vector product of their sum with **B** (see Appendix B), the total force on the length l is:

$$\mathbf{F} = I\mathbf{l} \times \mathbf{B} \tag{6.1}$$

If the wire is straight, all the elements experience forces in the same direction; the force on the whole wire is in this direction also, and is proportional to its length.

In the SI units which we are using, the force **F** should be given in newtons when we measure the length l in metres and the current I in amperes. This condition fixes the unit of flux density **B** in terms of the ampere which will be defined in section 6·6. This unit of flux density is called the weber per square metre, or Tesla.

Another well-known unit of flux density is the gauss, which belongs to the c.g.s. electromagnetic system of units (see Appendix A) and is equal to 10^{-4} Tesla.

6·2   **The couple on a coil**

Equation **(6.1)** was obtained as an expression for the force exerted by a uniform magnetic field on a straight wire, whose length is described by the vector l. Looking at the argument again, however, we see that it is equally valid for a piece of wire which is not straight, provided that l is the vector representing the straight line drawn from one end of the wire to the other. This is not the same as the length, if the wire is bent. The total force **F** tending to make the bent wire move sideways is made up of forces on the various elements of the wire which, not being parallel to each other, experience forces in different directions.

In the special case of a wire bent into a closed coil, the ends are close together and **l** defined in this way is zero: there is therefore no force tending to make the coil move bodily in any direction. There may, however, be forces in different directions on the various parts of the coil, giving it a tendency to rotate.

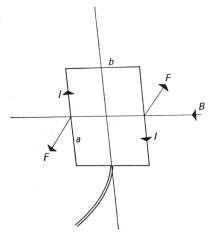

Figure 13.   Couple on a rectangular coil.

Let us first consider a rectangular coil lying in the plane of **B** (see figure 13). The sides which lie parallel to **B** experience no force as the vector product of two parallel vectors is zero, but the sides of length a, which are perpendicular to **B**, experience forces of magnitude IaB in opposite directions. Since these are acting at perpendicular distances $\frac{1}{2}b$ from the axis which has been drawn through the centre of the coil at right-angles to **B**, they together constitute a couple about this axis of magnitude

$$G = IabB = IAB \tag{6.2}$$

where A is the area of the coil.

The student is invited to confirm that IAB gives the couple on a coil of area A, whatever its shape, provided it lies in one plane and this plane contains **B**.

If **B** is in some other direction the effects of its components may be discussed separately:

First, the component perpendicular to the plane of the coil will give rise only to forces which tend to expand or contract the coil in its own plane.

Second, a component of **B** along the line we have called the axis would lead to forces giving no couple about this axis; instead they would constitute a couple about a perpendicular axis in the plane of the coil.

Third, the component $B_{\perp}$ in the plane of the coil, at right-angles to the axis, causes a couple $IAB_{\perp}$.

Thus there are two alternative ways of looking at the problem; we may either choose an axis in the plane of the coil and say that the couple about that axis is equal to IA times the component of **B** at right-angles to that axis, in the plane of the coil. Alternatively we may resolve a flux density **B** which makes an angle θ with the normal to the coil into two components: $B \cos \theta$ perpendicular to the coil, giving no couple, and $B \sin \theta$ in the plane of the coil, giving a couple $IAB \sin \theta$ about an axis perpendicular to **B**.

6·3   **The force between two parallel wires**

When currents are flowing in the same direction along two parallel wires, the wires are attracted towards each other. This force of attraction was discussed in Chapter 4 in terms of the diminished electrostatic repulsion of the moving electrons; in fact it formed the basis for our study of electromagnetic forces and magnetic fields. Now, in calculating the forces on current-carrying wires in given magnetic fields, we must take another look at our starting-point. We already know, from equation (5.4), that an infinite straight wire carrying a current I_1 produces at a distance a from itself a magnetic field:

$$H = \frac{I_1}{2\pi a}$$

at right-angles to its length. We also know from equation (6.1) that a second wire, carrying a current I_2 at right-angles to a magnetic flux density B, experiences a force per unit length equal to $I_2 B$. If B is due to the first wire it must be put equal to:

$$\mu_0 H = \mu_0 \frac{I_1}{2\pi a};$$

then the force per unit length on the second wire is:

$$F = \mu_0 \frac{I_1 I_2}{2\pi a} \tag{6.3}$$

towards the first wire if they are parallel.

It is interesting to note that although the force on an individual electron, due to a single electron, is inversely proportional to the square of their distance apart (as is the force on a piece of wire due to a single electron in the other wire), the force on a piece of one wire due to an infinite length of the other wire is inversely proportional to their distance apart (*not* the square of the distance). This is because, as a increases, the distance of the more remote electrons increases less rapidly than a and the factor sin θ in $\mathbf{u} \times \mathbf{r}$ increases (see figure 8). The net effect is that an infinitely long wire exerts on a single element of the other wire a total force depending inversely on the first power of the distance a.

In equation (6.3), I_1 and I_2 must be taken as the currents in a chosen direction along the wires; if the currents are in the same direction, I_1 and I_2 are both positive or both negative, thus giving a positive F which represents a force of attraction. However, if the currents are in opposite directions, $I_1 I_2$ is negative, and the negative F represents a force of repulsion.

6·4   Sign conventions and rules

6·4·1   *Sign convention in a vector product*

It is convenient at this point to review what we have already stated, without

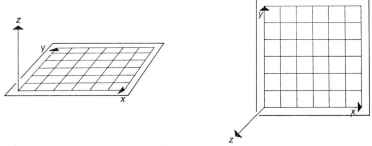

Figure 14.   Right-handed systems of axes.

emphasis, about the relative directions of currents, magnetic fields and forces. Everything is contained in the vector equations:

$$\mathbf{H} = \frac{1}{4\pi} \frac{I}{r^3} \, \delta\mathbf{s} \times \mathbf{r} \qquad\qquad (4.19)$$

$$\mathbf{B} = \mu_0 \mathbf{H}$$

and $\mathbf{F} = I\mathbf{l} \times \mathbf{B},$ (6.1)

provided we remember the sign convention included in the vector products: this convention is implicit in the expressions of Appendix B for the components of a vector product and may be summed up as follows: if vectors $\mathbf{x}$ and $\mathbf{y}$ are drawn respectively along the conventional x and y axes of a piece of graph-paper (figure 14) then their vector product $\mathbf{x} \times \mathbf{y}$ lies along a z axis drawn upwards from the paper. x, y and z axes arranged in this way form a right-handed set, characterized by the fact that the positive y axis is obtained from the positive x axis by a 90° rotation in the sense which is clockwise to an observer looking along the positive z axis.

4·2 *The corkscrew rule*

The direction of the magnetic field round a current element may be obtained by applying this convention to the vector product $\delta s \times r$. The result is that the magnetic field is in the direction of rotation of an ordinary corkscrew which is imagined as moving in the direction of the current. Since this fact is at least as easy to remember as the sign convention for vector products, it is often known and used as the 'corkscrew rule'.

4·3 *The direction of the force*

In order to ascertain the direction of the force on a current in a magnetic field, one must either use a set of right-handed axes for evaluating the vector product, or one must employ some equivalent mnemonic.

Mnemonics using the fingers of the hands for giving the direction of the magnetic field, and that of the force, are described in most elementary text-books. The author, however, finding difficulty in remembering the allocation of the fingers, as well as of the hands, prefers to remember the corkscrew rule and the fact that parallel currents attract each other. Together, these two facts provide a simple recipe for the direction of the force on a current in a specified magnetic field. For if the field is replaced by another current parallel to the first, in a position such that according to the corkscrew rule it would produce the specified field, then the force on the first current is towards the second. In addition, since he knows the convention for vector products, he tries to remember that the necessary products are $\delta s \times r$ and $I \times B$. To make it simpler, we can make the currents the vectors instead of the lengths, replacing $I\delta s$ by δsI and Il by lI; then the vector products are $I \times r$ and $I \times B$, both with I first.

4·4 *Polar and axial vectors*

It is important in this discussion to distinguish between fact and convention. Convention comes in whenever we consider the interaction of a current with a magnetic field. However, the force resulting from the interaction of a current with a current is fact; for even if we use a magnetic field to help calculate it, the answer will be the same whichever convention is used for vector products, since the convention is used twice. If left-handed axes were used for defining vector products, then our equations would give the same relations between forces and currents, but with magnetic fields and flux densities all having opposite signs. A similar element of convention arises when angular velocities, angular momenta and torques are represented by vectors. These vectors are known as axial vectors, to distinguish them from polar vectors like those of position and force which have a sense directly representing a physical reality.

To conclude this section, the student is advised to choose, according to his own tastes, a pair of facts or mnemonics which he can remember and use to deal with those problems in which signs and senses matter.

51   Forces due to magnetic fields

6·5   The force between two current elements

In Chapter 4 we discussed the force between two moving charged particles, and showed how the force due to their motion could be described in terms of a magnetic field created by one of them; the other experiencing a force as it moved through the magnetic field. We then calculated the magnetic field of a group of charges moving in such a way that they constituted a current element; and in a similar way the force exerted by a magnetic field on a current element was found.

The force on one current element, due to another, could be calculated by considering the force on one moving charge due to another, which is summarized in equations (4.14) and (4.15), and calculated in Appendix C; replacing each charge by a current element would then give the result which we shall now obtain by using the formulae already derived for current elements.

A single current element, acting as source, produces a magnetic field at a distance r which is given by equation (4.19); we call the current in this element I, and its length ds_1 so that it produces a magnetic field:

$$\mathbf{H} = \frac{1}{4\pi} \frac{I_1}{r^3} \, ds_1 \times r$$

In this field a second current element, of length ds_2, carrying a current I_2, experiences a force:

$$\mathbf{F}_2 = I_2 \, \mathbf{ds}_2 \times \mathbf{B}$$

$$= \frac{\mu_0}{4\pi} \frac{I_1 I_2}{r^3} \, \mathbf{ds}_2 \times (\mathbf{ds}_1 \times \mathbf{r}) \tag{6.4}$$

Considering the elements in reverse order and remembering that the vector from element 2 to element 1 is $(-\mathbf{r})$, we find that the force exerted by element 2 on element 1 is:

$$\mathbf{F}_1 = -\frac{\mu_0}{4\pi} \frac{I_1 I_2}{r^3} \, \mathbf{ds}_1 \times (\mathbf{ds}_2 \times \mathbf{r}) \tag{6.5}$$

These expressions contain triple vector products, which require some examination: when $\mathbf{ds}_1$ and $\mathbf{ds}_2$ are parallel to each other and at right-angles to $\mathbf{r}$, the triple products will be equal, and parallel to $\mathbf{r}$. The elements therefore experience equal and opposite forces along the line joining them (as we saw they must do from the simple arguments of Chapter 4).

However, in the general case, when $\mathbf{ds}_1$ and $\mathbf{ds}_2$ are not parallel, the forces are not equal and opposite; they are not even in the same direction, since $\mathbf{F}_2$ is perpendicular to $\mathbf{ds}_2$, and $\mathbf{F}_1$ is perpendicular to $\mathbf{ds}_1$. The clearest case of forces which do not balance each other is that with $\mathbf{ds}_1$ parallel to $\mathbf{r}$, and $\mathbf{ds}_2$ perpendicular to $\mathbf{r}$. In this case $\mathbf{F}_1$ is perpendicular to $\mathbf{r}$ and to $\mathbf{ds}_1$ $\left(= \dfrac{\mu_0}{4\pi} \dfrac{I_1 I_2}{r^2} \, ds_1 ds_2 \right)$; but $\mathbf{F}_2$ contains the vector product of two parallel vectors, and is therefore zero. This fact of current elements exerting forces on each other which are not equal and opposite, would appear to contradict Newton's third law: that action and reaction are equal and opposite. So they are, for realizable objects. When we build up a realizable object, such as a complete circuit, from a number of current elements, it is always found that the difference between the action and reaction has vanished. The physical explanation of this paradox is that the moving charges in a pair of isolated, non-parallel, current elements would have an electromagnetic field in which energy was moving from place to place, carrying with it momentum at a rate to balance the inequality of the forces on the current elements. However, real circuits carrying steady currents have electromagnetic fields which do not carry off momentum, so the difference between action and reaction must vanish as we perform the integration which takes us from isolated elements to complete circuits.

If it were not for the inconvenience of this paradox, equation (6.4) would be used more frequently as the basis for calculating the forces between currents, and this chapter would have started with it.

6·6 Definition of the ampere

6·1 In Chapter 2, we mentioned that electric charges could be measured in terms of a unit called the coulomb, which was the charge carried by a current of one ampere flowing for one second. Thus having introduced the ampere in this way and assumed that it would turn out to be definable in terms of some reproducible standard, we have used it throughout our discussion of magnetic fields and the forces which they exert on currents.

Now the point has been reached at which we can justify this assumption and replace our provisional ampere by a properly defined one. Since in equations (6.3) and (6.4) expressions have been obtained for the forces between two current elements and between two parallel current-carrying wires, we have an opportunity to define a unit of current as that which, when flowing through two wires in a given configuration, causes them to exert a specified force on each other. In other words the unit of current is correlated with the unit of force when a value for the constant μ_0 in equations (6.3) and (6.4) is chosen.

6·2 We are at liberty to choose a value for μ_0, since it was introduced in equation (4.12) as an abbreviation for $\dfrac{1}{\varepsilon_0 c^2}$, and ε_0 was at the beginning of Chapter 2 given the numerical value 8.85×10^{-12} in anticipation of the present section defining the ampere. The choice which leads to this value of ε_0 is:

$$\mu_0 = 4\pi \times 10^{-7} \text{ newton ampere}^{-2} \text{ or henry m.}^{-1}$$

The ampere may thus be considered as being defined as that current which, flowing in both of two long parallel wires one metre apart, causes them to exert on each other forces of 2×10^{-7} newtons per unit length. For current elements parallel to each other, one metre apart on a line at right-angles to their length, the force is $10^{-7} \, ds_1 ds_2$ newtons. Integration for other configurations, e.g. for the torque between two coils, gives related numerical factors.

6·6·3   This choice of μ_0 had the initial advantage of making the ampere one tenth of the cgs electromagnetic unit of current. The latter unit provided the theoretical basis for the definition in 1908 of the international ampere as the current which, flowing for one second in a suitable electrolytic cell, caused a specified amount of silver to be deposited on the cathode. This was meant to provide an accurately reproducible substandard equal to one tenth of a cgs electromagnetic unit of current; actually it was 150 parts per million too small. It survived until 1948, by which time it had become possible to use the electromagnetic definition with reproducibility at least equal to that which could be provided by the electrolytic standard.

6·6·4   Reverting for a moment to the value of ε_0, we see that this should be given by:

$$\varepsilon_0 = \frac{1}{\mu_0 c^2}$$

$$= \frac{10^7}{4\pi c^2}$$

If c is taken as 3×10^8 m. sec.$^{-1}$, this gives:

$$\varepsilon_0 = \frac{1}{(36\pi \times 10^9)} \quad \text{(in farad m.}^{-1}\text{)}$$

But if we use the best available value of c, namely $2\cdot99793 \times 10^8$ m. sec.$^{-1}$, we get:

$$\varepsilon_0 = 8\cdot85416 \times 10^{-12} \text{ farad m.}^{-1}$$

6·7   Motion of a charged particle in a magnetic field

6·7·1   *Radius of curvature of path*

An account of electromagnetism must overlap at many points with what is commonly taught as atomic physics. We have already drawn heavily on the atomic nature of matter and of electric charge to describe the fundamental electromagnetic effects; now we must point out an electromagnetic effect which is of great practical importance for research in atomic and nuclear physics.

We have seen in Chapter 4 [equation (4.15)] that a particle carrying electric charge q, moving with velocity $\mathbf{v}$ through a magnetic field with flux density $\mathbf{B}$, experiences a force:

$$\mathbf{F} = q\,\mathbf{v} \times \mathbf{B}$$

This force is at right-angles to the motion of the particle and it therefore gradually causes the particle to change direction. However, since force is defined as rate of change of momentum, the particle must acquire, in a short time δt, a sideways component of momentum equal to:

$$F\delta t = qvB\delta t$$

where $\mathbf{B}$ and $\mathbf{v}$ are perpendicular to each other and have magnitudes B and v. When combined with the longitudinal momentum mv, this sideways component causes the direction to change by a small angle given by:

$$\delta\theta = \frac{qvB\delta t}{mv} = \frac{qB\delta t}{m} \tag{6.6}$$

Here m is the mass of the particle, its ordinary mass if the velocity is small, or its relativistic mass if the velocity is large enough for the difference to matter. If we write the ordinary mass, or rest-mass, as m_0, the mass at velocity v is given by:

$$m = m_0\gamma = \frac{m_0}{\sqrt{\left(1-\dfrac{v^2}{c^2}\right)}} \tag{6.7}$$

but it is not really necessary to bother about this difference for the moment.

Equation (6.6) shows that the particle is changing its direction of motion at a uniform rate given by:

$$\frac{\delta\theta}{\delta t} = \frac{qB}{m}$$

In the limit of small intervals of time, $\dfrac{\delta\theta}{\delta t}$ just becomes the angular velocity $\dfrac{d\theta}{dt}$ with which the vector $\mathbf{v}$ is rotating.

Since the force on the particle is perpendicular to its velocity, no work is done by the force and the kinetic energy of the particle cannot change; this means that v, the magnitude of $\mathbf{v}$, must remain constant. If a particle is moving with constant velocity v in a direction changing at a uniform rate $\dfrac{d\theta}{dt}$, it is describing a circle of radius $\dfrac{v}{d\theta/dt}$, which in this case is $\dfrac{mv}{qB}$, where mv is the momentum of the particle for which the symbol p is normally used. Therefore, we may summarize our conclusions by saying that a charged particle moving in a uniform magnetic field describes a circular orbit of radius:

$$R = \frac{p}{qB} \tag{6.8}$$

If the magnetic field is not uniform then the path at any point is an arc of a circle of radius given by the same expression.

·2   *Application to atomic physics*

In elementary experiments in atomic physics and also in the ordinary television tube, beams of charged particles are accelerated to definite velocities by passing them through an electrostatic potential difference, and then bending them through an angle inversely proportional to their momentum by means of a magnetic field. This field usually extends far enough to make them follow only a small arc of a circular path, after which they emerge in a straight line tangential to the circle.

However, in accelerating machines like the cyclotron (figure 15), particles are made to follow a complete circular path, going round many times. The magnetic

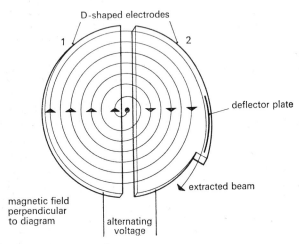

Figure 15. The cyclotron.

field in this case serves as a guide and the particles are accelerated by a potential difference applied between a pair of electrodes. The time taken for each orbit is $\frac{2\pi R}{v}$ which equation **(6.8)** gives as $\frac{2\pi m}{qB}$. This is independent of R and v as long as v is small enough to allow us to neglect the relativistic factor in m. In the cyclotron therefore the charged particles circulate inside two hollow D-shaped accelerating electrodes, which are connected to a source of alternating e.m.f. If a particle crosses from electrode 1 to electrode 2 when the voltage between them is at a peak in the direction needed for acceleration, it will go round half of its circular orbit while the voltage is changing to a peak in the opposite direction, and then it will be accelerated again as it crosses back to electrode 1. The remaining part of the orbit inside electrode 1 allows the alternating voltage again to build up to a peak in the first direction. Thus if the accelerating voltage is made to alternate, with one cycle taking a time $\frac{2m}{qB}$, the particle will go on and on being accelerated twice in every orbit. As the velocity of the particle increases, the radius of the orbit will increase. Therefore the energy attainable from a cyclotron depends on the radius of the region over which the magnetic field extends.

Cyclotrons based on this simple principle are used for accelerating protons (nuclei of hydrogen atoms) up to energies equivalent to those obtained by a single passage through a potential difference of about 12 million volts. The energy of such a proton is said to be 12 million electron-volts (MeV.). In order to extend this energy up to about 750 MeV., we need not only very large electromagnets, but also a device for keeping the alternating voltage in step with the rotating particles as their mass undergoes relativistic increase. This is done in the synchrocyclotron by gradually decreasing the frequency of the alternating voltage, so that one group of particles can keep in step with it and continue to be accelerated by it. The frequency is then increased again, so that in decreasing once more it can accelerate another group of particles. By modulating the frequency in this way, it is possible to obtain bursts of particles at regular intervals of the order of 1/50 second.

However, synchrocyclotrons (sometimes called frequency-modulated cyclotrons) have large magnets and limited potentialities; for even higher energies it is necessary to use a synchrotron, in which both the frequency and the magnetic field are varied. The variations are synchronized in such a way as to keep R constant as v increases. This allows the use of a ring magnet, which produces field only over the orbit of chosen radius. The inside of the orbit can be empty space, so that field can be applied to an orbit of radius about 100 metres with a magnet whose total weight is less than that required for a synchrocyclotron of radius a few metres.

Chapter 7
Electromagnetic induction

7·1   **Induced e.m.f.**

7·1·1 It has already been shown that a wire carrying a current experiences a force when it is situated in a magnetic field: it follows that work must be done when a wire carrying a current moves through a magnetic field. But since energy cannot be created or destroyed, any such work done must be provided from some source. This source is usually the agency which is causing current to flow in the wire, and we shall now discuss the means by which energy is extracted from it:

We shall take, as a simple example, a coil of area A mounted so that it can rotate about an axis in its own plane. This is placed in a uniform magnetic field with flux density **B**, in a direction perpendicular to the axis, making an angle θ with the normal to the coil. We have seen in section 6·2 that such a coil experiences a couple $IAB \sin \theta$.

Now suppose that external forces cause the coil to rotate with angular velocity $\dfrac{d\theta}{dt}$ against the couple; the external forces will be doing work at a rate equal to the product of the couple and the angular velocity, namely:

$$IAB \sin \theta \, \frac{d\theta}{dt}$$

This work will be supplied to the circuit, where it will help to maintain the current I in its passage through the ohmic resistance of the circuit. If the current I is to carry energy away from the coil at this rate, there must be an electromotive force V in the coil, aiding the flow of current, and supplying work at a rate IV.

V is called an induced e.m.f. and if the rate at which external forces do work on the coil is made equal to the rate at which the current carries away energy from it we get:

$$IV = IAB \sin \theta \, \frac{d\theta}{dt}$$

$$\therefore \ V = AB \sin \theta \, \frac{d\theta}{dt}$$

$$= -AB \frac{d}{dt} (\cos \theta)$$

$$= -\frac{dN}{dt} \qquad \qquad (7.1)$$

where:

$$N = AB \cos \theta$$

which is the area of the coil multiplied by the component of flux density at right-angles to the coil. This is called the magnetic flux through the coil. Since flux density has been measured in terms of a unit called the weber per square metre, it follows that the unit of flux is the weber. In cases where the flux density B is not uniform, the flux is equal to the surface integral (see section 8·2) of B, i.e. $\int B \, . \, dS$ over the area of the circuit. If the circuit contains a coil with more than one turn the integral must be taken over every turn, so N will be equal to the flux through the coil times the number of turns, plus the flux through the rest of the circuit.

1·2   It has thus been shown that when the flux N through a circuit is changing at a rate $\frac{dN}{dt}$, there will be in it an induced e.m.f. V numerically equal to $\frac{dN}{dt}$ More. general arguments show that this holds for any circuit whatever the reason for the change of flux through it: whether the change is in the flux density, in the area of the circuit, or in their relative orientations, there will be an induced e.m.f. in the circuit equal to the rate of change of magnetic flux through it.

If the units are such that I is in amperes, N is in webers and B is in webers per m.², then the couple will be in newton-metres and the rate of doing work in newton-metres per second (metres parallel to the force for work, cf. metres perpendicular to the force for couple). But since a newton-metre is a joule and a joule per second is called a watt, the rate of doing work will be given in watts; this is the same as the rate of supply of energy from the source if V is in volts. It can thus be seen that the rate of change of flux in webers per second is equal to the induced e.m.f. in volts.

The significance of the negative sign in equation (7.1) deserves some consideration. If the angle θ had been carefully defined, we should have found that N represented the flux measured in the same direction as that which would be produced by the current I, and that V was an e.m.f. measured in the same direction as that needed to cause the current I. With N and V defined in these senses, the negative sign shows that the induced e.m.f. tends to oppose the change of flux. That is, if the flux through the circuit is increased, the induced e.m.f. will create currents which tend to decrease the flux. This is known as Lenz's law.

7·2   Self-inductance

2·1   *The nature of self-inductance*

In deducing that an e.m.f. of magnitude $\frac{dN}{dt}$ is induced in a circuit by a rate of change of magnetic flux $\dfrac{dN}{dt}$, we were thinking in terms of a flux due to agencies outside the circuit. However, the argument holds whatever the source of the flux, so flux due to a current in the circuit itself is in no way exempt. Indeed it was seen in Chapter 5 that a current I flowing in the circuit must produce at each point in its neighbourhood a magnetic field proportional to I. The constant of proportionality for each point depends on the geometry (perfect proportionality breaks down when ferromagnetic materials are present: see Chapter 8). It follows that as the line integral of magnetic field round the wire is equal to I [equation (5.10)], the field must be distributed so that it gives a flux through the circuit. If a wire is in the form of a circuit, 'round the wire' means 'through the circuit' (figure 16). We may therefore assume that a given circuit will possess

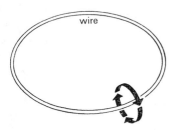

wire

Figure 16. 'Round the wire is through the circuit.'

a geometrical constant L, such that a current I round it sets up a magnetic flux LI through it.

As the current I changes, the flux will change and there will be set up in the circuit an induced e.m.f.:

$$V = -\frac{dN}{dt} = -L\frac{dI}{dt} \tag{7.2}$$

The constant L, which is called the self-inductance, may thus be defined in either of two ways; as the magnetic flux through the circuit per unit current, or as the induced e.m.f. per unit rate of change of current. Its unit is thus the weber per ampere or volt per (ampere per second), which has a special name, the henry. This is rather too large for most purposes, so millihenries and microhenries are commonly used.

Equation (7.2) provides a very direct indication of the need for the negative sign in equation (7.1), because without this negative sign equation (7.2) would mean that any increase in current, represented by a positive $\dfrac{dI}{dt}$, led to a positive induced e.m.f., which would tend to increase the current even further. The law of conservation of energy, as well as common sense, show that currents cannot in this way cause themselves to increase beyond limit. But with the negative sign, the equations describe an effect which tends to slow down any variation in current.

Any current flowing in any wire produces some magnetic field, so even the simplest circuit must have some self-inductance. However, in section 7·1 we pointed out that a coil could make a large contribution to the flux through a circuit, simply because the flux inside the coil passes once through the circuit for every turn in the coil. When the coil is also the source of the flux, its contribution is even greater, because the flux density itself is proportional to the number of turns. In fact, for coils in which the turns are much closer than their dimensions, the self-inductance must be proportional to the square of the number of turns. From this and other similar facts we can infer that it is reasonable to speak of the self-inductance of components, as well as of complete circuits. To a first approximation, which is usually sufficient, the self-inductance of a circuit is the sum of the self-inductances of its components. Occasionally, however, it may be necessary to allow for the fact that one component is creating a magnetic field which is causing an induced e.m.f. in another part of the circuit. In this situation, the mutual inductance of the two components must be included; but that will be dealt with in section 7·3.

The procedure used to calculate the self-inductance of an actual coil or of some other arrangement of wires, is in principle simple. The Biot-Savart law [equation (4.20)] is used to give the magnetic field at a point inside the circuit, due to unit current flowing in one short element of the wire; then we integrate over the whole wire to get the total field at that point. Finally we insert a factor μ_0 to convert the field to flux density, and integrate over the whole area enclosed by the coil to get the flux N, which is equal to the self-inductance. It is also possible to perform the integrations in the reverse order, by first calculating the flux through the coil due to one element of the wire, and then integrating along the wire to get the flux due to the whole of it. The idea is straightforward, but the integrations are often so cumbersome that it is necessary to make approximations. We may also have to make simplifying approximations concerning the distribution of current within a wire. In our first example, a solenoid, we go even farther than this and treat the current through a sequence of turns as a uniform cylindrical sheet of rotating current.

7·2·2 *Solenoid*

To calculate the self-inductance of a long solenoid, we use equation (5.9), which gave the field at a point on the axis inside the solenoid as mI, where m is the number of turns per unit length. This expression was derived by an integration which treated the turns as parts of a continuous sheet of current, a very good approximation so long as the spacing between the turns is much smaller than their radii.

It must now be shown that for a sufficiently long solenoid, the field inside it is uniform. If the length of the solenoid is so great that variations of field near the ends can be ignored, then the line integral of H around a closed path down the axis and back along a loop some distance outside the solenoid is simply lH where l is the length. (It is assumed that the field outside is negligible.)

Putting $H = mI$, this gives:

$$\oint \mathbf{H} . \mathbf{ds} = lmI \tag{7.3}$$

lm is the total number of turns, and the line integral of H has to be equal to the current times the number of turns, so this provides a check that our expression for H is correct. Indeed, if we had not wished to demonstrate the direct method of calculating magnetic fields, equation (7.3) could have been used to derive the expression for the field on the axis, rather than to check it. Equation (7.3) applies to any path through the solenoid, including paths which go through it straight, parallel to the axis but displaced from it. Since the length of such a path inside the solenoid is l, which is so great that variations along the length can be neglected, the field at points off the axis must have the same value $H = mI$ as it has on the axis.

Now that the field is known to be uniform over the inside of a solenoid, the flux through each turn can immediately be written as:

$$BA = \mu_0 mI\pi a^2$$

where a is the radius. This flux goes through the circuit ml times, so the self-inductance is:

$$L = \frac{BAml}{I} = \pi\mu_0 m^2 a^2 l \tag{7.4}$$

When we are dealing with any real solenoid, it is necessary to allow for the finite length; a first-order correction may be calculated by the student, using equation (5.8) for the field near the ends.

7·2·3 Stored energy

In the derivation of equation (7.1), we used the fact that when an induced e.m.f. V is present in a circuit carrying a current I, the source of the induced e.m.f. is feeding energy into the circuit at a rate VI. This must be true whether the induced e.m.f. results from motion of a coil in a magnetic field, as in section 7·1, or from a flux changing for some other reason. The statement that energy is being transferred from the source of flux to the circuit at a rate VI is equivalent to saying that the source of current in the circuit is providing energy at a rate $(-VI)$. In cases where V is negative (with respect to I) so that $(-VI)$ is positive, this latter statement is the more direct description of what is happening.

If the negative V is due to a positive $\dfrac{dN}{dt}$ in a coil rotating in a magnetic field, the energy provided by the source of the current is used in doing work against the external forces resisting the rotation of the coil. But if the negative V is due to a positive $\dfrac{dN}{dt}$ in a self-inductance, there are no external mechanical forces against which work can be done. We must therefore ask where the energy which leaves the source can go. If there is no machinery for converting it into mechanical work and it is definitely energy additional to that which is being dissipated in the ohmic resistance of the circuit, it must be stored somewhere. Since the flux N is increasing, it can reasonably be concluded that the disappearing energy is somehow stored in the flux, because when the flux decreases again, it can feed its stored energy back into the circuit.

In a self-inductance, the flux N is proportional to the current, and the amount of energy stored in the flux, when a current I is flowing, can be calculated: with $N = LI$, the rate at which the source provides energy becomes:

$$-VI = I\frac{dN}{dt} = LI\frac{dI}{dt}$$

During the whole process of increasing the current from zero to a value I, the total amount of energy provided by the source must therefore be:

$$\int LI\frac{dI}{dt}\,dt = \int LI\,dI = \tfrac{1}{2}LI^2 \tag{7.5}$$

We must therefore conclude that whenever a current I is flowing in a self-inductance L, an amount of energy $\tfrac{1}{2}LI^2$ is stored.

If energy is to be stored in a self-inductance, it must be associated with the magnetic field; we can calculate how, and how much, by considering the long solenoid whose self-inductance we have already calculated. When a current I flows in the long solenoid of equation (7.4), the stored energy is:

$$\tfrac{1}{2}LI^2 = \tfrac{1}{2}\pi\mu_0 m^2 a^2 lI^2$$

This energy is stored in a magnetic field of intensity $H = mI$, which extends over a volume $\pi a^2 l$ (the total volume enclosed by the solenoid). Hence the energy stored per unit volume of field is:

$$\frac{\tfrac{1}{2}\pi\mu_0 m^2 a^2 lI^2}{\pi a^2 l} = \tfrac{1}{2}\mu_0 m^2 I^2 = \tfrac{1}{2}\mu_0 H^2 \tag{7.6}$$

Although this calculation has used an ideal solenoid, with uniform field inside and no field outside, it does yield a universally valid expression for the energy per unit volume of a magnetic field. It can be shown by more general arguments that wherever there is a magnetic field H, there is a density of stored energy equal to $\tfrac{1}{2}\mu_0 H^2$.

2·4   *Internal self-inductance*

4·1   So far we have considered wires carrying currents which produce magnetic flux around the wires: if the wires form a circuit, the amount of the flux which goes through it determines the self-inductance of the circuit. But as we discussed in section 5·5, there may be some flux inside the wires and the effect of this must be taken into account when calculating the self-inductance. When thinking of flux inside a wire, it is difficult to see how to be exact in asking, or answering, the question 'does it go through the circuit?' It is therefore necessary to find a way of calculating its contribution without asking this question. Such a way is provided by the calculations of stored energy which have just been carried out. If the magnetic field at all points inside the wire is known, when a given current is flowing, then the stored energy can be calculated by integrating $\frac{1}{2}\mu_0 H^2$ over the volume of the wire, and saying that this must be equal to $\frac{1}{2}L_i I^2$, where L_i is the contribution of the energy in the wire to its self-inductance. For convenience this contribution is called the internal self-inductance, to distinguish it from the total which is obtained by adding the two.

7·2·4·2   We now have to examine the question of how the current is distributed in the wire, and the extent to which the magnetic field is dependent on the distribution of current. In Chapter 5 it was shown that the magnetic field at a distance r from a long straight wire (assumed thin) was $\dfrac{i}{2\pi r}$, when current i was flowing in the wire, and thus that the line integral of **H** along any path round the wire was equal to i. (For convenience we are now using symbols i and r for I and a.) The line integral is in general $\oint \mathbf{H}.\mathbf{dl}$, but for a circular path around the wire and concentric with it, **H** is parallel to **dl** all the way round, and the line integral is just the (uniform) magnitude of **H** multiplied by the circumference of the path. This fact was used for the simple case in which all the current flows along a thin wire passing through the centre of the path; but it holds for any case in which the field has the same magnitude at all points on the path. If the current is flowing along a thick cylindrical wire and is distributed with cylindrical symmetry (i.e. there is no difference between one side of the wire and another), all points on the circular path will be equivalent, and the above condition is satisfied. Thus the equation

$$H = \frac{i}{2\pi r} \qquad \text{(7.7), equivalent to (5.4)}$$

holds for the field due to any cylindrically-symmetric distribution of current, with its axis at a distance r from the point of observation. There is, however, one proviso; if the wire should be made so thick that it extends beyond the point of observation (or if the point of observation should be made so close to the axis that it lies inside the wire), only those parts of the current which lie inside the radius r contribute to the line integral and to H.

7·2·4·3   We have thus arrived at two important conclusions which apply to cylindrically-symmetric distributions of current.

(i) The magnetic field at a distance r from the axis depends on the total current flowing within the cylinder of radius r, and not on how the current is distributed inside this cylinder. This current may be concentrated on the axis, or near the outside, or it may be uniformly distributed, but so long as it is cylindrically symmetric its external field is the same.

(ii) A cylindrically-symmetric sheath of current creates no magnetic field inside itself.

In practice we meet a whole range of cylindrically-symmetric distributions of current, but they all lie between two extremes which are especially easy to handle:

(a) High-frequency alternating currents which flow entirely in a thin skin close to the surface of a wire, and not in the body of the wire. This effect is discussed in Chapter 13. Because of this superficial current flow, the 'external' self-inductance, due to fields and fluxes outside the wire, is sometimes called the high-frequency self-inductance. It is calculated on the assumption that the current is flowing in an infinitely thin cylindrical sheet on the surface of the wire.

(b) Steady or slowly-varying currents, in which the current density is uniform over the cross-section of the wire. Such a uniform distribution of current produces the same field outside as would the same current flowing in a thin skin on the surface, and also produces some flux inside the wire. The self-inductance for steady or slowly-varying currents is therefore equal to the high-frequency self-inductance, plus an internal self-inductance which we shall now calculate.

4·4   If a total current I is flowing in a wire of radius a, and is uniformly distributed over its cross-sectional area πa^2, the current inside a smaller radius r is:

$$i = I\frac{r^2}{a^2}$$

From equation (7.7), the magnetic field at radius r in such a wire is therefore:

$$H = I\frac{r}{2\pi a^2},$$

and the stored energy per unit volume at radius r is:

$$\tfrac{1}{2}\mu_0 H^2 = \mu_0\, I^2\frac{r^2}{8\pi^2 a^4}$$

The stored energy in unit length of the wire, between radii r and $r+dr$ is therefore:

$$\tfrac{1}{2}\,\mu_0 H^2 \times 2\pi r dr = \mu_0\, I^2\frac{r^3}{4\pi a^4}\,dr,$$

and the total stored energy per unit length is the integral of this from 0 to a:

$$\int_o^a \mu_0\, I^2\frac{r^3}{4\pi a^4}\,dr$$

$$= \frac{\mu_0 I^2}{4\pi a^4}\left[\frac{1}{4}r^4\right]_o^a$$

$$= \frac{\mu_0 I^2}{16\pi}$$

We have seen that this must be equal to $\frac{1}{2}L_iI^2$, so the internal self-inductance per unit length is:

$$L_i = \frac{\mu_0}{8\pi} \tag{7.8}$$

This is a constant quantity, independent of the radius of the wire, and valid for any round wire carrying a uniformly-distributed current. If we insert the numerical value of μ_0, $4\pi \times 10^{-7}$ henry per metre, L_i takes an even simpler form:

$$L_i = \tfrac{1}{2} \times 10^{-7} \text{ henry per metre}$$

$$= \tfrac{1}{2} \times 10^{-9} \text{ henry per centimetre.}$$

The internal self-inductance of a round wire carrying a uniformly-distributed current is therefore one half a millimicrohenry per centimetre length, or $1/20$ microhenry per metre.

This is the amount which must be added to the high-frequency self-inductance to get the low-frequency or steady-current self-inductance. At intermediate frequencies, the current partially penetrates the interior of the wire, so intermediate values of self-inductance are observed. A real piece of apparatus therefore has its largest self-inductance at very low frequencies; the value of L decreases as frequency increases, gradually approaching its minimum value as the frequency rises and the depth of penetration (sometimes called the skin depth) decreases.

In practice, the internal self-inductance adds little to the impedance of a wire, because if the frequency ω is small enough for L_i to be effective, ωL_i is usually much less than the resistance R.

7·2·5 Coaxial cable

For the first exact calculation of a self-inductance, we shall derive the high-frequency self-inductance of a coaxial cable, in which the current flows along an inner cylindrical wire (on the skin of it in the high-frequency limit, which will be considered first), and back along a cylindrical sheath coaxial with the wire. When the inner and outer currents are perfectly balanced, e.g. by short-circuiting the conductors at one end, they produce no magnetic field outside, since the net current passing through any path outside the cable is zero. In fact the net flux through the circuit is just the flux N around the inner wire, in the space between it and the sheath. In order to clarify the meaning of 'through the circuit' in the case of a coaxial cable, the circuit may be considered as being made up of a large number of wedge-shaped radial sections of the cable, connected in parallel; each such section constitutes a one-turn coil, through which the flux N passes. All the sections therefore have the same e.m.f. $\left(-\dfrac{dN}{dt}\right)$ induced in them, and since they are connected in parallel to form the cable, this is the e.m.f. induced in it; the value of N per unit current is the self-inductance of the cable. If the inner and outer conductors have radii b and a respectively, the self-inductance per unit length is therefore:

$$L = \frac{N}{I} = \frac{1}{I}\int_b^a \mu_0 \times H \, dr = \int_b^a \mu_0 \frac{dr}{2\pi r}$$

$$= \frac{\mu_0}{2\pi} \log_e\left(\frac{a}{b}\right) \tag{7.9}$$

The student is invited to check that the same result is obtained by calculating the total stored energy from the volume integral of $\tfrac{1}{2}\mu_0 H^2$ over the space between the conductors, and equating it to $\tfrac{1}{2}LI^2$.

This expression, (7.9), is for the high-frequency limit, in which all the current on the inner wire is flowing in a thin skin on its surface. As we have already seen, slowly-varying currents are uniformly distributed through the inner wire. This leads to an extra, internal self-inductance $L_i = \dfrac{\mu_0}{8\pi}$ per unit length, giving a total self-inductance per unit length

$$L_{\text{TOTAL}} = \frac{\mu_0}{2\pi}\left\{\log_e\left(\frac{a}{b}\right) + \frac{1}{4}\right\} \tag{7.10}$$

7·6 Self-inductance of parallel wires

The self-inductance of a pair of long straight cylindrical wires, lying parallel to each other, is relatively easy to calculate on the basis of 'flux through the circuit'. Initially, the internal self-inductance of the wires is ignored, i.e. the high-frequency self-inductance is calculated; also it is assumed that the distance a between the wires is so much greater than their radius b that the cylindrical symmetry of the current distribution in each wire is not disturbed. Under these conditions, the field due to a current I in one wire, at a distance r from its axis, is $H = \dfrac{I}{2\pi r}$. This gives a flux through the space between the wires, per unit length, equal to

$$\int_b^a \frac{\mu_0 I}{2\pi r}\, dr = \frac{\mu_0 I}{2\pi}\log_e\left(\frac{a}{b}\right)$$

But when the current I is flowing along one wire and back along the other, the two wires produce equal fluxes in the same direction, through the space between them. The self-inductance, given by the flux through the circuit per unit current in it, is therefore twice the flux due to each wire. For unit length of the wires this is:

$$L = \frac{\mu_0}{\pi} \log_e\left(\frac{a}{b}\right) \tag{7.11}$$

At low frequencies, we must add the internal self-inductance, which is $L_i = \dfrac{\mu_0}{8\pi}$

per unit length of each wire. For unit length of the pair of wires the total self inductance at low frequencies is therefore:

$$L_{\text{TOTAL}} = \frac{\mu_0}{\pi} \left\{ \log_e \left(\frac{a}{b} \right) + \frac{1}{4} \right\}$$ (7.12)

7·2·7 Self-inductance of a single-turn coil

For a single-turn circular coil made out of round wire whose radius is much less than that of the coil, the high-frequency self-inductance may be calculated by integrating the flux through the coil due to each element of wire, on the assumption that the wire is straight enough for the current to be spread uniformly over its surface. The integration is not particularly easy, but yields the result:

$$L = \mu_0 a \left\{ \log_e \left(\frac{8a}{b} \right) - 2 \right\},$$ (7.13)

where a is the radius of the coil, and b that of the wire.

For low frequencies an internal self-inductance equal to $\frac{\mu_0}{8\pi}$ times the circumference $2\pi a$ must be added; this gives a total:

$$L_{\text{TOTAL}} = \mu_0 a \left\{ \log_e \left(\frac{8a}{b} \right) - \frac{7}{4} \right\}$$ (7.14)

Formulae equivalent to this and others quoted in this chapter are listed in handbooks such as F. E. Terman, *Radio Engineer's Handbook*, McGraw Hill 1943.

7·2·8 Magnitudes

To obtain a feeling for the magnitudes of these quantities we now insert some values. For a coil with $\frac{a}{b} = 200$, the high-frequency self-inductance is about 6·8 microhenries per metre radius; this means that a single turn of 15 cm. radius made of wire with radius 0·75 mm., has a high-frequency self-inductance very close to 1 microhenry. The internal self-inductance of this coil would be 0·015 microhenries. Therefore the low-frequency self-inductance would be about 1·0 microhenries, with 0·05 coming from the internal self-inductance which that length of wire would have in any moderately tightly bent configuration, and from the shape of the coil.

In comparison with this, the formula for high-frequency self-inductance of a pair of parallel wires gives 1 microhenry per metre length when $\log_e \left(\frac{a}{b} \right) = 2·5$

e.g. for wires of radius 0·75 mm. separated by 9 mm. But in this case the low frequency value is 10 per cent larger than the high-frequency figure.

7·3   Mutual inductance

7·3·1   *The nature of mutual inductance*

·3·1·1   When two circuits are placed close together, a current in one may give rise to a flux through the other; hence a changing current in one may cause an induced e.m.f. in the other. Expressing this quantitatively, a current I_1 in the primary circuit causes a flux:

$$N = MI_1$$

through the secondary circuit. When I_1 changes, there is induced in the secondary circuit an e.m.f.:

$$V_2 = -\frac{dN}{dt} = -M\frac{dI_1}{dt} \tag{7.15}$$

The mutual inductance M may therefore be defined either as the flux through the secondary circuit due to unit current in the primary, or as the e.m.f. induced in the secondary circuit per unit rate of change of current in the primary. In practice we do not need to bother about the negative sign in equation (7.15) unless the relative senses of the primary and secondary circuits are specified. The commonest example of a mutual inductance is that provided by a pair of coils, wound on a common former, one connected to the primary circuit and the other to the secondary circuit.

3·1·2   Before continuing our discussion of mutual inductance any further, we must show that it deserves its name. The term 'mutual' is often used loosely, to describe a one-way interaction between two circuits or parts of a circuit; for example, the name mutual conductance is given to what is more aptly called the transconductance of a triode valve, namely the rate of change of anode current with grid voltage. But there is nothing mutual about this, just as there is nothing mutual about a friendship in which one party does all the giving. 'Mutual' in its true sense is defined in the *Oxford Concise English Dictionary*, as 'felt, or done, by each to(ward) the other'. We shall now show that mutual inductance is mutual in this sense.

Suppose that a steady current I_2 is flowing in circuit 2, while a current i_1 is increased from zero to I_1 in the circuit 1. At any moment, there will be an induced e.m.f. in circuit 2 given by equation (7.15) as $-M\frac{di_1}{dt}$, which causes the source of the current I_2 to feed energy into the magnetic field at a rate $I_2M\frac{di_1}{dt}$. During the whole process of increasing i_1 from 0 to I_1, the total energy stored up in this way is:

$$W = \int I_2M\frac{di_1}{dt}\,dt = \int_0^{I_1} I_2M\,di_1 = MI_1I_2 \tag{7.16}$$

Similarly suppose that a steady current I_1 is flowing in circuit 1, while the current i_2 in circuit 2 is increased from zero to I_2. If a changing i_2 causes an induced e.m.f. $-M'\dfrac{di_2}{dt}$ in circuit 1, the energy stored in the field while i_2 rises from 0 to I_2 comes by a similar argument to $M'I_1I_2$.

However, the state of the system is defined by the magnitudes of the currents in the two circuits, and the stored energy must be a function of these two currents, not of the process by which they were built up. If this were not so, one could extract energy continually by repeatedly setting up the currents in the way which required least energy, and decreasing them in the way which yielded most. Therefore, when the state defined by I_1 and I_2 is set up by increasing I_2 last, the energy stored $M'I_1I_2$ must be the same as the energy MI_1I_2 which is stored when I_1 is increased last. So M' must be the same as M; i.e. the e.m.f. induced in circuit 1 per unit rate of change of current in circuit 2 is the same as the e.m.f. induced in circuit 2 per unit rate of change of current in circuit 1. In other words, the mutual inductance really is mutual.

7·3·1·3   Mutual inductance cannot exist without some self-inductance, and we must now take a first look at the relations between them. First of all, they are both measured in terms of flux per unit current, or of induced e.m.f. per unit rate of change of current, so the henry serves as a unit for mutual as well as for self-inductance.

Next in the case of a single circuit containing two coils in series both carrying current I, the flux through coil 1 will be L_1I due to its own self-inductance L_1, plus MI due to its mutual inductance M with coil 2; and the flux through coil 2 will similarly be L_2I due to L_2, plus MI due to coil 1. The total flux through the circuit is therefore $(L_1+L_2+2M)I$, and the effective self-inductance of the two coils is

$$L_{max} = L_1+L_2+2M \tag{7.17}$$

If coil 2 is turned through 180°, the flux due to the mutual inductance will now oppose that due to L_1 and L_2, and the effective self-inductance of the whole arrangement will be

$$L_{min} = L_1+L_2-2M \tag{7.18}$$

Since no arrangement of wires and coils can have a negative, or even a zero, self-inductance, equation (7.18) shows that $2M$ must be less than L_1+L_2, i.e. that two coils cannot have a mutual inductance as large as the mean of their self-inductances. We shall see in Chapter 15, from consideration of alternating currents, that it is possible to set an even more stringent limit on M: it cannot be greater than the geometric mean of L_1 and L_2.

7·3·2   *Coil wound round solenoid*

It is relatively simple to calculate the mutual inductance of a pair of coils in which one is a long solenoid, and the other is a few turns wound tightly around the central part of the solenoid. If the solenoid is regarded as the primary, unit current in it will produce a flux through it equal to $\mu_0 \pi a^2 m$, where m is the number of turns per unit length on the solenoid, and a its radius. This flux will cut the secondary circuit n times, where n is the total number of turns on the secondary coil (*not* the number per unit length). The mutual inductance is therefore

$$M = \mu_0 \pi a^2 mn \qquad (7.19)$$

This expression is much more accurate than the corresponding expression (7.4) for the self-inductance of a long solenoid, because in calculating the latter we assumed that the flux through the end turns was the same as that through the middle ones; but in deriving the mutual inductance, we have merely ignored the effect of the ends on the field at the centre. The correction for this effect is much smaller and more easily calculated. For this reason, it is fairly easy to construct a standard mutual inductance whose value is known from its dimensions.

This section is slender for the simple reason that later in the book a whole chapter (15) is devoted to the most important type of mutual inductance, the transformer.

7·4   Induced electric field

4·1   *Evaluation of induced electric field*

We have seen in the preceding sections that if a circular loop of wire is placed in a region where the magnetic field is changing, there is induced in it an e.m.f. numerically equal to the rate of change of magnetic flux through it. It does not matter how much flux is changing outside, nor which part of the inside contains the changing flux N. The e.m.f. V is still equal to $-\dfrac{dN}{dt}$.

Thus when an electric charge q moves round the wire, under the influence of the induced e.m.f., the work done on it by the changing flux is:

$$W = qV = -q\frac{dN}{dt} \qquad (7.20)$$

Now we can reduce the wire to a path in space along which electric charges can move. The work done on a charge q in moving right round this path must still be given by equation (7.20); but this work can no longer be regarded as a single unit of energy presented to the charge as it moves through an induced e.m.f., for it represents a steady increase in the energy stored as kinetic energy of the charged particle as it goes round the path. If the kinetic energy of the charged particle is increasing, then according to the ideas of Chapter 2 it must be experiencing an electric field E. Using the same notation as in section 2·3, the total work done on it as it goes round the path is q times the line integral $\int E . ds$. The induced e.m.f. V which appears in the wire is therefore equal to the line integral of an induced electric field which is effective even in the absence of a wire. Equation **(7.1)** is thus a particular case of a more general equation.

$$\oint \mathbf{E} \cdot \mathbf{ds} = -\frac{dN}{dt} \qquad (7.21)$$

Here $\oint$ means line integral round a closed path encircling the flux N.

If the wire or path is circular, and arranged symmetrically about the changing flux, we have an arrangement of flux and wire with a cylindrical symmetry similar, but converse, to that which was demanded in section 7·2 for currents flowing through paths along which line integrals of H were calculated. In this case there is nothing to distinguish one point on the path from another, as E is of uniform magnitude E all the way round. The line integral $\int \mathbf{E} \cdot \mathbf{ds}$ round it is therefore equal to E times the circumference. Hence if the path has radius a, E must be given by:

$$E = \frac{V}{2\pi a} = -\frac{1}{2\pi a}\frac{dN}{dt} \qquad (7.22)$$

7·4·2   Electric potential in a changing magnetic field

At this point we must reconsider a statement which was emphasized in section 2·3: that the line integral $\int \mathbf{E} \cdot \mathbf{ds}$ along a closed path is zero. This applies for the static systems discussed in Chapter 2 which contain no sources of energy and only limited amounts of stored electrostatic energy; but the presence of a changing magnetic field, from which energy can be drawn, allows $\int \mathbf{E} \cdot \mathbf{ds}$ around a closed path to have a definite non-zero value. For this reason, the idea of electrostatic potential in space has only limited usefulness when changing magnetic fields are present; it has to be considered as a multi-valued function, the different values being obtained according to whether the point is approached by a path that goes once, twice, or n times round the changing magnetic flux.

7·4·3   The betatron

There is a special machine which depends for its operation on the existence of an induced electric field in free space. This is the betatron, in which electrons are accelerated in a circular path by a changing magnetic flux through the path. A large flux is arranged to go through the orbit which the electrons are to follow, but the electromagnet producing it is designed to produce also a magnetic field in the neighbourhood of the desired orbit. Hence by suitably contriving the relative magnitudes, it can be arranged that this field is always of the correct strength for keeping the electrons in the chosen orbit. This is reasoned as follows: suppose the orbit is of radius R, with an average flux density B through it at time t; the flux through the orbit is $\pi R^2 B$, so the electric field acting on an electron is, by equation **(7.22)**:

$$E = -\frac{\pi R^2}{2\pi R}\frac{dB}{dt} = -\frac{R}{2}\frac{dB}{dt}$$

The electron has a charge $q = -e$, so the force on it is $-eE$ in the direction of motion, and this may be put equal to its rate of increase of momentum. In time δt, the momentum increases by

$$\delta p = \frac{R}{2} \frac{dB}{dt} \delta t$$
$$= \tfrac{1}{2} e R \delta B \tag{7.23}$$

if δB is the amount by which B changes in time δt.

This means that if the electrons start with zero momentum when B is zero, the momentum then increases in proportion to B, so that it is always equal to $\tfrac{1}{2}eRB$.

However, to hold the electrons in an orbit of radius R, when their momentum is p, requires a flux density B' at the orbit. This is given by equation **(6.8)** as:

$$B' = \frac{p}{eR}$$

If we now put:

$$p = \frac{1}{2}eRB,$$

this gives:

$$B' = \frac{\tfrac{1}{2}eRB}{eR} = \frac{1}{2}B.$$

Thus if electrons are to be kept in an orbit of constant radius in a betatron, the flux density at the orbit should be half the mean flux density over the area inside the orbit.

This conclusion, and the arguments on which it is based, are valid for relativistic as well as for low velocities.

5   Maxwell's equation for curl E

Taking equation **(7.21)** and applying it to the flux through a small area δS, we get:

$$\frac{\oint \mathbf{E}.\mathbf{ds}}{\delta S} = -\frac{dB}{dt} \cos \phi \tag{7.24}$$

where ϕ is the angle between $\mathbf{B}$ and the normal to the area δS. This equation is closely similar to equation **(5.13)**, with $-\dfrac{dB}{dt}$ replacing j and $\oint \mathbf{E}.\mathbf{ds}$ replacing $\oint \mathbf{H}.\mathbf{ds}$. As in the earlier case, the line integral divided by the area may be reduced to a vector normal to the plane which gives the largest value of line integral per unit area, and this vector is called the curl of the vector whose line integral it includes. Thus equation **(7.24)** leads us to the vector equation:

$$\text{curl } \mathbf{E} = -\frac{d\mathbf{B}}{dt} \qquad (7.25)$$

This is the second of Maxwell's four equations; equation (5.15) was the first.

If equation (7.25) is chosen to be the basic relation between changing magnetic field and induced electric field, we need to avoid several pitfalls. In spite of its appearance, this equation does not relate the value of **E** at a point to the rate of change of **B** at that point; it gives not **E**, but curl **E**, which specifies how much **E** differs from place to place. The actual value of **E** at one place will depend upon the value of curl **E** over a region extending from it to a place where **E** is known. This may be demonstrated by considering the question 'how is an electron passing beside an electromagnet affected by an increasing magnetic field between the poles of the magnet?' The answer is that it experiences a force due to the changing flux beside it, even if there is no magnetic field at the point where it is. In figure

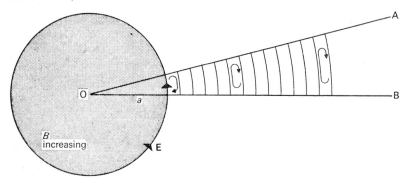

Figure 17. Curl **E** = 0 does not imply **E** = 0.

17, if the changing flux is concentrated inside a circle of radius a, the line integral of **E** round it is $-\dfrac{dN}{dt}$ and the electric field at the edge of it is $-\dfrac{1}{2\pi a}\dfrac{dN}{dt}$, so with no changing flux outside the circle, curl **E** will be zero everywhere outside. This means that the line integral of **E** round any small closed path will be zero. In particular it will be zero round the closed paths shown in figure 17, which are each bounded by two radii OA and OB and by two arcs of concentric circles. From the symmetry of the problem, the radial sides of one of these paths can contribute nothing to the line integral, therefore the two arcs must contribute equal and opposite amounts. Since the lengths of the arcs are proportional to their distances from O, this means that the electric field must fall off inversely as the distance from O; so at distance r, E must be equal to $-\dfrac{1}{2\pi r}\dfrac{dN}{dt}$. In this way the knowledge that curl **E** is zero allows us to correlate the tangential electric fields in different places. It does not allow us to say that E is zero anywhere until we have checked whether it is non-zero somewhere close by.

The velocity of electromagnetic radiation

In free space, where there are no free charges and no currents, Maxwell's equation **(5.15)** becomes

$$\text{curl } \mathbf{H} = \frac{d\mathbf{D}}{dt} \tag{7.26}$$

which we may rewrite in the form

$$\text{curl } \mathbf{B} = \mu_0 \, \varepsilon_0 \, \frac{d\mathbf{E}}{dt}. \tag{7.27}$$

This relates the space-dependence of $\mathbf{B}$ with the time dependence of $\mathbf{E}$, just as the second Maxwell equation **(7.25)** relates the space-dependence of $\mathbf{E}$ with the time-dependence of $\mathbf{B}$.

We may solve these equations simultaneously to find separate equations for $\mathbf{E}$ and $\mathbf{B}$. This is most easily done by taking the curl of equation **(7.25)** to get:

$$\text{curl (curl } \mathbf{E}) = -\text{curl}\left(\frac{d\mathbf{B}}{dt}\right) = -\frac{d}{dt}(\text{curl } \mathbf{B})$$

which by substituting from equation **(7.27)** gives

$$\text{curl (curl } \mathbf{E}) = -\mu_0\varepsilon_0 \frac{d^2\mathbf{E}}{dt^2}. \tag{7.28}$$

Subject to the condition that $\mathbf{E}$ has zero divergence (see chapter 9 and Appendix B), curl (curl $\mathbf{E}$) reduces to

$$\left(\frac{d^2}{dx^2} + \frac{d^2}{dy^2} + \frac{d^2}{dz^2}\right)\mathbf{E}$$

which may be written in terms of the vector operator ∇ as $\nabla^2\mathbf{E}$. Equation **(7.28)** thus gives us, as a general solution for $\mathbf{E}$ in free space

$$\nabla^2\mathbf{E} = -\mu_0\varepsilon_0 \frac{d^2}{dt^2}\mathbf{E} \tag{7.29}$$

This is a wave equation, and is a starting-point for discussing the transmission of electromagnetic radiation. A similar equation holds for $\mathbf{B}$, and modified equations apply when free charges or polarisable materials are present.

Discussion of electromagnetic radiation is beyond the scope of this book and for it we must refer the student to more advanced texts. We must, however, draw the reader's attention to the velocity implied by equation **(7.29)**: the general equation for a wave of velocity v is

$$\nabla^2\mathbf{E} = -\frac{1}{v^2}\frac{d^2}{dt^2}\mathbf{E},$$

so we deduce that the velocity of electromagnetic waves in free space is $(\mu_0\varepsilon_0)^{-\frac{1}{2}}$. But the definitions of μ_0 and ε_0 (see sections 4.5 and 6.6) have already fixed this quantity as equal to c, the velocity of light used in the relativistic transformations by which magnetic and electric forces are related. We have thus shown that the "velocity of light" used in relativistic transformations is the same as the velocity of electromagnetic radiation in free space. The velocity used in relativistic transformations is really an upper limit to the possible velocity of transmission of signals and also to the relative velocity of movement of objects with non-zero mass.

So what we have done is to prove that electromagnetic radiation offers the fastest possible transmission of signals, and sets a limit to physical velocities, and that light being a type of electromagnetic radiation, the name "velocity of light" is appropriate to the quantity used in relativity theory. This is a particularly good illustration of the nature of electromagnetic theory as a web of interlocking arguments characterised more by self-consistency than by the merit of any particular single line of reasoning.

Chapter 8
Magnetostatics

1 *Magnetic moment*

Magnetism and electricity started life as separate subjects, for it was, and indeed still is, possible to study the interactions of the objects called magnets without reference to electricity. Magnets were known and studied long before a basis had been laid for the electromagnetism described in the preceding chapters of this book. Many detailed descriptions have been written of the properties of magnets, and of the magnetic properties of the earth which led to the use of the magnetic compass for navigation; we shall not write another, but shall assume some general knowledge of what a magnet is.

When the forces which magnets exert upon each other are studied, it is found that each magnet has a characteristic direction, which is called its direction of magnetization. It also has a characteristic strength specified by its magnetic moment, which describes the magnitude of its interactions with other magnets. By virtue of its magnetic moment, a magnet produces around itself a field which can exert forces or couples on other magnets, by virtue of *their* magnetic moments. Experimentally, this field is found to be of the same nature as the magnetic field which has already been used to describe the effects of moving electric charges: a coil carrying an electric current exerts forces upon magnets as if it were itself a magnet; and conversely, magnets exert forces upon it as if it were a magnet.

We shall account for this correspondence qualitatively, by suggesting that the properties of magnets result from the presence in them of rotating or spinning electric charges. These charges are, of course, on an atomic scale, and they will be discussed in more detail later.

2 *Comparison of couples created by magnet and coil*

The next task is to examine quantitatively the way in which magnets and current-carrying coils are equivalent. The magnetic moment m of a magnet is defined as the couple per unit **B** which it experiences when placed in an external magnetic field at right-angles to its direction of magnetization. m may be considered as the magnitude of a vector magnetic moment **m** which is directed along the direction of magnetization; then if the couple is described by a vector **G** whose direction and magnitude give respectively the axis and the magnitude of the couple, **G** is just the vector product of **m** with the flux density **B**:

$$G = m \times B \qquad \text{(8.1)}$$

But if the magnitudes G, m and B are used, G is given by

$$G = mB \sin\theta \qquad \text{(8.2)}$$

where θ is the angle which the field makes with the direction of magnetization of the magnet.

Equation **(8.2)** may be compared with the expression $IAB \sin\theta$, given in section 6·2 for the couple on a coil of area A carrying a current I in a magnetic field with flux density B. The comparison shows that the couple on a coil is equivalent to that on a magnet with magnetic moment:

$$m = IA \qquad \text{(8.3)}$$

So far as couples are concerned, this equivalence must apply also to the fields produced by the coil and the magnet. Newton's third law states that when a real object* exerts a force on another object, it must itself experience a reaction equal and opposite to the force. This applies to couples just as it applies to forces of translation. If a coil and a magnet which are related by equation **(8.3)** experience equal couples when they are placed (separately) at a given position in the field of a distant magnet, this source magnet must experience equal couples of reaction in the two cases. In other words, the coil and the magnet must exert equal couples on the distant magnet; but to do so, they must be creating, in the neighbourhood of the distant magnet, equal magnetic fields.

8·1·3   *Comparison between fields of magnet and coil*

If the fields of the magnet and the coil are to be compared in more detail, we must examine the rules for calculating the magnetic field of a magnet. The descriptions give us the following rule: a magnet of moment m behaves as if it contained two equal and opposite magnetic poles of strength $\pm p$, separated by a distance l such that $\mu_0 m = pl$. The magnetic field of the magnet is as if these poles produced magnetic field by an inverse square law of the form:

$$H = \pm \frac{p}{kr^2}, \qquad \text{(8.4)}$$

$+$ for the positive (or North) pole and $-$ for the negative (or South) pole.

---

* We specify a real object, for the reasons discussed in section 6·5: Newton's third law is not satisfied by the conventional expressions for the force between two current elements; but integration over a complete circuit gives a result which does satisfy it. A corresponding discussion may arise apropos of moving charged particles, between which equations **(4.15)** and **(4.10)** imply unbalanced magnetic forces. The student may like to check for himself that the *total* forces exerted on each other by two moving charged particles do nevertheless satisfy Newton's third law, provided that the same frame of reference is used for calculating them; this point is implicit in the argument of Appendix C.

According to this recipe, the total magnetic field at a distant point on the axis of the magnet, distant x from the centre of the magnet, would be:

$$H = \frac{p}{k\left(x-\dfrac{l}{2}\right)^2} - \frac{p}{k\left(x+\dfrac{l}{2}\right)^2}$$

$$= \frac{p}{kx^2}\left\{\left(1-\frac{l}{2x}\right)^{-2} - \left(1+\frac{l}{2x}\right)^{-2}\right\}$$

If $\dfrac{l}{2x}$ is sufficiently small, the curly bracket may be reduced by the binomial expansion to $\dfrac{2l}{x}$, and H becomes:

$$H = \frac{2pl}{kx^3} = \frac{2\mu_0 m}{kx^3} \tag{8.5}$$

Now we can refer to equation (5.7), which gave the magnetic field of a coil of area A at a point distant x on the axis as:

$$H = \frac{IA}{2\pi x^3} \tag{5.7}$$

Equations (8.5) and (5.7) contain the same dependence on x, and are compatible with equation (8.3) if k is put equal to $4\pi\mu_0$. So if we want to think of these hypothetical magnetic poles, the condition for making the conclusions compatible with electromagnetism is that a pole p_1 should be said to produce at a distance r a magnetic field:

$$H = \frac{p_1}{4\pi\mu_0 r^2}, \text{ with } B = \frac{p_1}{4\pi r^2} \tag{8.6}$$

If this field is to exert a couple $m_2 B$ on a second magnet of moment m_2, placed at this distance r, we must say that the poles $\pm p_2$ of this magnet experience forces due to p_1 of magnitude:

$$F = \pm \frac{p_1 p_2}{4\pi\mu_0 r^2} \tag{8.7}$$

This law of force appears to have such a pleasing similarity to that for the electrostatic force between two electric charges that it has sometimes in the past been used as the basis for an elementary treatment of magnetism, but here it is presented merely as a link between electromagnetism and the simple theory of permanent magnets, which may be called magnetostatics. The possible existence of free magnetic poles, "magnetic monopoles", has been the subject of much theoretical and experimental investigation: Dirac in 1930 showed by an argument based on phases of wave-functions that the magnitude of any magnetic monopole must be a multiple of a unit p, related to the electronic charge by the relation $pe = \dfrac{e^2}{hc} \approx \dfrac{1}{137}$. Experimental searches for magnetic monopoles which might be produced in particle accelerators or by the cosmic rays have so far given negative or inconclusive results. We therefore continue our discussion of magnetic fields under the simplifying assumption that they are entirely due to moving charges and that monopoles do not exist; but we note that elementary electromagnetic theory could be quite naturally extended to include magnetic monopoles if these were to be observed.

8·1·4 *Equivalent magnetic shell*

Referring back to equation (8.3) we see that a coil carrying a current I experiences forces identical to those which would be felt by a magnet occupying the same area, if it had magnetic moment per unit area equal to I. The direction of the magnetic moment must be normal to the plane of the coil. If this is studied in more detail, the coil can be regarded as being made up of a network of little coils, each carrying a circulating current I, so that the outside wire of the network is equivalent to our coil, and the inner links all carry zero total current. Then each little coil may be declared equivalent to a short magnet of magnetic moment equal to I times its area, in a direction normal to the plane of its area; it has to be short, if we are not to run into difficulties over its effective parts not being in the same place as the coil. The magnets do not have to be distinct, and they can be run together into a uniformly magnetized sheet, which is everywhere magnetized at right-angles to its surface, with magnetic moment per unit area equal to I. Such a sheet is called a magnetic shell and if it is cut to have the same shape and area as the coil, it will have the same magnetic field, even at nearby points. This idea of the equivalent magnetic shell, though somewhat artificial, has served well as a convenient basis for calculating the magnetic fields of coils, as was mentioned in Chapter 5; it is found in a prominent place in many useful textbooks.

8·2 Gauss's law

In order to study one of the rules which magnetic flux in free space must obey, whatever its source, we start by using as source a single current element, the magnetic field of which was described in Chapter 4 and then used as a basis for the calculations of Chapter 5.

First let us calculate the total magnetic flux leaving the closed surface S in figure 18, as a result of a current I flowing in an element δs of a wire. If the magnetic field at the point P is $\mathbf{H}$, the flux over a small piece of the surface, near P, is the scalar product **B.δS**, where **B** is μ_0**H** and **δS** is a vector normal to the piece of the surface with length equal to its area. The flux over the whole of the surface S will be obtained by taking the sum of the scalar products **B.δS** for all the small pieces δS which together make up the whole area of S.

If we now look at the configuration of magnetic field due to the current element, we see from equation **(4.20)** that the field at P is of magnitude $\dfrac{I\delta s \, \sin \theta}{4\pi r^2}$, in a direction along the circumference of a circle through P with centre on the axis of the current element. Here r is the distance from P to the element δs, which may be anywhere on the wire, above or below the plane of the circle, and θ is the angle between the directions of the wire and of the line from δs to P. The circle through P is a line of constant r, so the magnetic field has the same magnitude all the way round the circle. In fact every circle with centre on the wire is a line

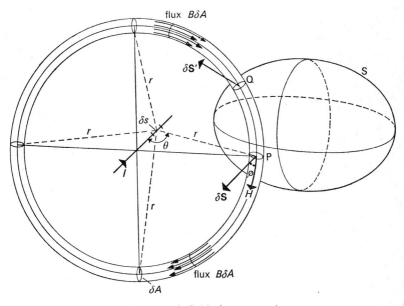

Figure 18.   Gauss's law for magnetic field of a current element.

of constant H, everywhere parallel to the vector **H** whose magnitude is H. So we may pierce our closed surface S with a whole series of mathematical circles, and especially circles of constant H drawn through the boundary of the little area δS, in such a way that they form a tube. This tube will have cross-sectional area

$\delta A = \delta S \cos \phi$, where ϕ is the angle between **H** and the normal to the surface; the same ϕ can now be used to rewrite the flux **B.δS** as $B\delta S \cos \phi$, or $B\delta A$. This means that a flux $B\delta A$ leaves S along the tube at P. If we go on round the tube, its cross-sectional area remains δA, and the same flux $B\delta A$ re-enters S through a small area $\delta S'$ around the point Q. The total flux leaving S through δS and $\delta S'$ together is therefore zero. Therefore the whole of S may be cut up into corresponding pairs of small areas like δS and $\delta S'$, through which equal fluxes leave and enter, and when the total flux leaving S is summed, this must be zero.

Thus the total magnetic flux leaving any closed surface, due to any current element, is zero; the current element is drawn outside the surface in figure 18, but it could have been inside without affecting the conclusion, so we really do mean *any* current element. This being so, we may use the fact that fields are superposable, and fluxes additive, to say that the magnetic field of a whole wire is equivalent to the superposed fields of all the elements of the wire. Each of these contributes zero to the flux leaving a closed surface and therefore no arrangement of currents or wires can create a field in which the flux leaving any closed surface is non-zero.

Extending this even further by saying that all magnetic fields are due to moving electric charges, we obtain Gauss's law, that the magnetic flux leaving a closed surface is always zero. In vector language, this is written:

$$\int B.dS = 0 \qquad\qquad (8.8)$$

This is Gauss's law for magnetism; there is a corresponding law for electrostatics, which is discussed in section 9·1. There we show that the inverse square law of electrostatic forces inevitably leads us to conclude that the total electric flux out of a closed surface is numerically equal to the magnitude of the electric charge inside. The same reasoning may be applied to any force obeying an inverse square law; so if magnetism is imagined as a phenomenon governed by forces between magnetic poles, with forces obeying equation (8.7) which is an inverse square law, we may conclude that the total magnetic flux leaving any closed surface is numerically equal to the strength of any pole inside. Since magnetic poles are never found separately, there can be no isolated pole inside a closed surface, and thus the magnetic flux leaving it must be zero. This may be regarded as either an alternative or a principal proof of Gauss's law in magnetism.

8·3   **Maxwell's equation for div B**

Gauss's law, as it has been considered in the preceding section, tells us that the total magnetic flux leaving a closed surface is zero, whatever the shape and size of the surface. We now want to apply it to the surface of the small rectangular parallelepiped shown in figure 19. The sides AB, AD and AE are denoted by δx, δy and δz respectively, to remind us that they are small. The magnetic flux density in the neighbourhood of the parallelepiped is **B**, with components B_x, B_y and B_z in the direction of the three axes.

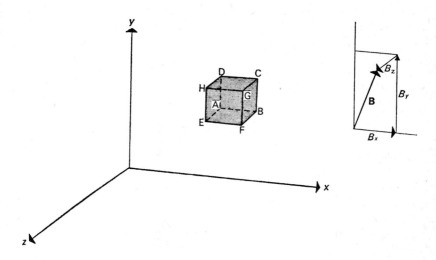

Figure 19. Components of flux density and flux.

The fluxes over the surfaces ADHE and BCGF will be given by their areas multiplied by the x-components of the flux density at the surfaces. The latter will not be exactly equal unless the field is uniform; if it is non-uniform, with B_x changing with x at a rate given by the partial differential coefficient $\dfrac{\partial B_x}{\partial x}$, B_x at the surface BCGF will be greater than B_x at the surface ADHE by an amount $\dfrac{\partial B_x}{\partial x}\,\delta x$. The areas are both equal to $\delta y \delta z$, and therefore the flux outward over BCGF exceeds the flux inward over ADHE by an amount $\dfrac{\partial B_x}{\partial x}\,\delta x \delta y \delta z$. This is the net outward flux provided by these two opposite surfaces, and it will be positive or negative according to the sign of $\dfrac{\partial B_x}{\partial x}$.

The flux over two surfaces ABFE and DCGH has no contribution from the components B_x and B_z which are parallel to the surfaces; the fluxes over them determined by the values of B_y at the two surfaces and these values differ by $\dfrac{\partial B_y}{\delta y}\,\delta y$. The area of each surface is $\delta x \delta z$, and we conclude that the net outward flux over the two surfaces is $\dfrac{\partial B_y}{\partial y}\, \delta y\, \delta x\, \delta z$.

By a corresponding argument, the two remaining surfaces, ABFE and DCGH, contribute a net outward flux equal to $\dfrac{\partial B_z}{\partial z}\, \delta z\, \delta y\, \delta x$.

The total outward flux over the closed surface made up of the outside of the parallelepiped is the sum of the sub-totals for these three pairs of opposite faces, which can be written as:

$$\text{Total flux} = \left(\frac{\partial B_x}{\partial x} + \frac{\partial B_y}{\partial y} + \frac{\partial B_z}{\partial z}\right) \delta x\, \delta y\, \delta z \tag{8.9}$$

This has to be zero, by Gauss's law, and as $\delta x \delta y \delta z$ is the volume of the parallelepiped, which though small is not zero, the quantity in the bracket is zero. This quantity is a function of the way in which **B** varies from place to place and is a derivative of **B** in the sense that it is obtained by differentiating the components of **B**, each along its own axis, and adding the results. It is thus a directionless quantity or scalar, obtained from the vector **B**. Any vector can be treated in this way and the number obtained is called its divergence. So the quantity in our bracket is the divergence of **B**, which is written as div **B**.

We thus conclude that Gauss's law, applied to a small volume, requires that:

$$\frac{\partial B_x}{\partial x} + \frac{\partial B_y}{\partial y} + \frac{\partial B_z}{\partial z} = 0,$$

which is written in the simplified form:

$$\text{div } \mathbf{B} = 0 \tag{8.10}$$

Equation (8.10) is sometimes referred to as the differential form of Gauss's law: it describes the limitation which Gauss's law imposes on the differential coefficients of the components of **B** at a single point.

If we wish to replace the particular argument above by a more general mathematical one, we may use Gauss's theorem (see Appendix C) to relate the two forms of Gauss's law.

In another context, equation (8.10) is called one of Maxwell's four equations; we have already met Maxwell's equations for curl **H** and curl **E** in equations (5.16) and (7.25), and we have one more to come [equation (9.7)].

8·4   **Magnetization**

8·4·1   *Source of magnetization*

We have already hinted that the properties of magnets depend upon the presence in them of electric charges which are rotating or spinning on an atomic scale.

In a strong permanent magnet a large number of atoms have to act cooperatively to give the observed total magnetic moment, one electron from each atom providing a small magnetic moment. This cooperation is obtained by aligning all the electron orbits so that their magnetic moments add. Materials in which this type of cooperative magnetization occurs are called ferromagnetic, since iron is the commonest example.

·2   Diamagnetism

Many materials, however, have magnetic properties without being ferromagnetic, and without having the possibility of maintaining any permanent magnetic dipole moment. If a bag of miscellaneous bits of wire, which are all good conductors of electricity, is placed in an increasing magnetic field, currents will be induced in the wires and by Lenz's law these will all be in the directions which tend to create magnetic flux opposing that which is being applied from the outside. In other words, the current flowing round a loop of wire will create a flux through the loop which partly cancels the applied flux; but outside the loop, the field due to the current will be in the same direction as the applied field. This is illustrated in figure 20 which shows the separate fields and the total field obtained by vector addition of the applied and induced fields. The latter shows that the effect of the induced current is to prevent flux from passing through the loop. In the bag of wires the induced currents will flow only while the applied magnetic field is increasing, for if it decreases again, they will flow in the opposite direction, always opposing change at points inside. The electrons spinning and rotating in ordinary matter have some properties in common with our bag of wires: they can by changing their motion simulate microscopic

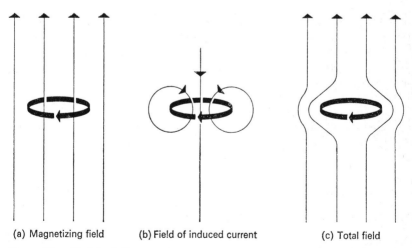

(a) Magnetizing field   (b) Field of induced current   (c) Total field

Figure 20.   Model of diamagnetism.

induced currents. But these currents, unlike those induced in wires, do not die away; once made by an increase in magnetic field they continue to discourage the penetration of magnetic flux through them. In fact the induced magnetic moment is proportional to the applied magnetic field, irrespective of the historical process by which the field was built up; it only returns to zero when the applied field becomes zero again.

All materials possess, to a greater or lesser extent, the possibility of acquiring an induced magnetic moment through this process. Those in which this process predominates over the others and which can produce magnetic moment in the opposite direction are called diamagnetic.

8·4·3 *Paramagnetism*

Certain types of atom contain electrons moving in orbits of such a nature that the atom has a permanent magnetic moment. But in material at ordinary temperatures, such atoms are subject to jostling by their neighbours: the thermal energy stored in the material is present as energy of vibration of the atoms about their average position. This thermal agitation will effectively alter the direction of the magnetic moment of any atom which may have one; so individual atomic

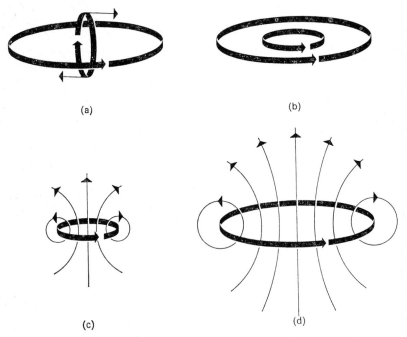

(a)   (b)

(c)   (d)

Figure 21.   Model of paramagnetism: alignment of an existing magnetic moment causes reinforcement of field along the axis.

magnetic moments will be oriented randomly. But when a magnetic field is applied to the material, each atomic magnetic moment will experience a couple which tends to align it in the direction of the magnetic field. How much alignment takes place depends on the strength of the magnetic field and on that of the thermal agitation which continually opposes the alignment. In many materials, the amount of alignment is proportional to the applied magnetic field and inversely proportional to the absolute temperature. However, in a specimen containing a large number of atoms, the components of the atomic magnetic moment at right-angles to the applied field cancel and the net effect of the alignment is that the specimen acquires a magnetic moment in the direction of the applied field.

This phenomenon of partial alignment of existing atomic magnetic moments, against the influence of thermal agitation, is known as paramagnetism. Its practical difference from diamagnetism lies in the direction of the magnetic moment acquired by the specimen. We can see this direction by reference to figure 21, in which we suppose the applied field to be due to a current in a large coil; the rotating charge which gives the atom its magnetic moment has been drawn to resemble a small coil carrying a current. We know from Chapter 4 that parallel currents tend to attract each other, while antiparallel currents tend to repel each other. Thus in figure 21(a) the atomic magnetic moment experiences a couple which in the absence of thermal agitation would make it take up the position of figure 21(b). When it is in this position, it has a magnetic field of the type shown in figure 21(c). Comparison with the field of the large coil (the applied field), drawn in figure 21(d), shows that the fields reinforce each other in the middle of the atom. This is the exact opposite of the situation drawn in figure 20 to represent diamagnetism.

4·4   Susceptibility

For both diamagnetic and paramagnetic materials, the total magnetic moment acquired by a specimen is proportional to the number of atoms in it, and also to the applied magnetic field. The material can therefore be said to have acquired a certain magnetic moment per unit volume, which is called magnetization. This is usually described by a vector $\mathbf{M}$, whose magnitude is the magnetic moment per unit volume, and whose direction is that which we called the 'direction of magnetization' in section 8·1. Since $\mathbf{M}$ is in the same direction as the applied magnetic field $\mathbf{H}$ and its magnitude is proportional to that of $\mathbf{H}$, the vectors are proportional as follows:

$$\mathbf{M} = \chi_m \mathbf{H} \tag{8.11}$$

The constant of proportionality χ_m is called the magnetic susceptibility of the material.

In a paramagnetic material the magnetization is said to be in the same direction as the magnetic field, so that χ_m is positive. As it has already been explained, the magnetization of such a material by a given applied field varies linearly with the reciprocal of the absolute temperature; it follows that paramagnetic susceptibilities are temperature-dependent in the same way.

A convention having been set for what is called positive magnetization, a diamagnetic material must be given a negative susceptibility. Such negative susceptibilities are found to be independent of temperature; they are observed in materials where all the electron orbits and spins are arranged so that their permanent magnetic moments cancel. This cancelling usually takes place between pairs of electrons in similar states; but when some electrons are unpaired, a permanent magnetic moment exists, and the paramagnetic effect is likely to swamp the natural diamagnetism of the more numerous paired electrons.

Paramagnetic and diamagnetic susceptibilities are extremely small. For example, at room temperature and one atmosphere pressure, oxygen is paramagnetic with $\chi_m = 0.21 \times 10^{-5}$ and silver is diamagnetic with $\chi_m = -2.6 \times 10^{-5}$.

8·5   Effect of magnetic material

8·5·1   *Effect of magnetization on the equations of the field*

8·5·1·1   In the preceding chapters we have developed a series of equations which describe the behaviour of electric and magnetic fields in free space, in which apart from the occasional object carrying an electric charge, or a wire carrying an electric current, the space has been assumed empty. So the results obtained are strictly valid only *in vacuo*; usually air is a sufficiently good approximation to vacuum, and the results can be applied to systems in air without noticeable error.

But when we consider solid materials, the applicability of our equations needs a careful review. We can start this review by examining the magnetization M in the light of equation (8.3). When this equation was discussed at the end of section 8·1, it was seen that the magnetic effects of a coil carrying a current *I* are the same as those of a similar-shaped magnetic shell with magnetic moment per unit area *I*. Reversing the argument, we may say that a magnet of area *A* and magnetic moment *MA* has the same effects as would a current of magnitude *M* circulating round the outside of the area *A*. Indeed, since the magnetization is due to rotating and spinning electric charges in the atoms, to describe it in terms of one large circulating current is no more far-fetched than to describe it as a magnetic moment per unit volume. The two are equivalent descriptions; each has its own sphere of usefulness, but neither has any fundamental superiority over the other.

8·5·1·2   Imagine a non-uniformly magnetized specimen, of which two small blocks are drawn in figure 22; if the mean magnetization over one of these blocks has x-component M_x, it will have x-component of magnetic moment per unit area equal to $M_x \delta x$. This component will therefore have the same magnetic effects as a current *I* circulating round its outer surface, where

$$I = M_x \delta x \tag{8.12}$$

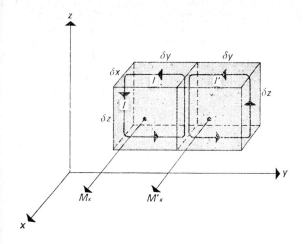

Figure 22. Magnetization as a circulating current.

The second block, touching the first but displaced relative to it by a distance δy, equal to the y-dimension of each block, will have a mean magnetization with x-component:

$$M'_x = M_x + \frac{\partial M_x}{\partial y}\,\delta y.$$

Therefore, the x-component of its magnetization will have the same effect as if it were replaced by a current I' circulating round its outer surface, where:

$$I' = M'_x \delta x = \left(M_x + \frac{\partial M_x}{\partial y}\,\delta y\right)\delta x \tag{8.13}$$

Equations (8.12) and (8.13) together show that the interface of the two blocks carries a net current in the direction of the z-axis (upwards in the figure), of magnitude:

$$I - I' = -\frac{\partial M_x}{\partial y}\,\delta x\,\delta y \tag{8.14}$$

This upward current will not really be localized in the mathematical common surface of the two blocks, for if the magnetization is varying smoothly with position, the current may be considered spread over an area $\delta x\,\delta y$ at right-angles to the z-axis. In this same area there will be another upward current, which is obtained by considering two similar blocks with their relative displacement along the x-axis, and their common surface parallel to the y–z plane; if we examine the y-component of magnetization in these blocks, we find that there is a net upward current of magnitude:

$$\frac{\partial M_y}{\partial x} \, \delta x \, \delta y$$

There is therefore a total upward current, in the area $\delta x \, \delta y$, of magnitude:

$$\left(\frac{\partial M_y}{\partial x} - \frac{\partial M_x}{\partial y} \right) \delta x \, \delta y$$

The total upward current per unit area may be put equal to the z-component of a current density $\mathbf{j}'$. The expression $\left(\frac{\partial M_y}{\partial x} - \frac{\partial M_x}{\partial y} \right)$ is just the z-component of the vector curl $\mathbf{M}$, so with the areas cancelling we have shown that the z-component of $\mathbf{j}'$ is equal to the z-component of curl $\mathbf{M}$. The corresponding relation holds for the other components, so the vectors themselves may be equated:

$$\text{curl } \mathbf{M} = \mathbf{j}' \tag{8.15}$$

$\mathbf{j}'$ is not an ordinary current density and it has been given a dash to distinguish it from $\mathbf{j}$, which is the density of flow of ordinary electric current. $\mathbf{j}'$ is a current density associated with the magnetization $\mathbf{M}$ and it may be considered as a real current whose effect is usually described by $\mathbf{M}$, or as an artificial alternative way of describing $\mathbf{M}$, or simply as an alternative description without statement of relative significance.

8·5·1·3   So having established that equation (8.15) relates two descriptions of the same phenomenon, we now refer back to equation (5.14) which relates magnetic field $\mathbf{H}$ to the current density creating it:

$$\text{curl } \mathbf{H} = \mathbf{j} \tag{5.14}$$

This relates the ordinary current density to the applied magnetic field $\mathbf{H}$, the field which describes the effect of sources other than the rotating atomic charges. It is apparent from the form of equations (8.15) and (5.14) that the combined effect of the magnetization and the flow of real current is described by adding the two equations:

$$\text{curl } (\mathbf{H} + \mathbf{M}) = (\mathbf{j} + \mathbf{j}') \tag{8.16}$$

Now we come to a definition: in a magnetic material $\mathbf{H}$ is the field excluding that due to magnetization; but the term flux density, with symbol $\mathbf{B}$, describes the total effect of real moving charges and magnetization together. In free space, $\mathbf{B}$ is equal to $\mu_0 \mathbf{H}$, but in a magnetized material, $\mathbf{B}$ is given by:

$$\mathbf{B} = \mu_0 (\mathbf{H} + \mathbf{M}) \tag{8.17}$$

At this point it might be helpful to mention that, while we have given $\mathbf{M}$ the same dimensions as $\mathbf{H}$, some books (including the first edition of this) give $\mathbf{M}$ the same dimensions as **B**. The transposition to this convention could be made by writing $\dfrac{\mathbf{M}}{\mu_0}$ for **M** and $\dfrac{m}{\mu_0}$ for m. This can be done relatively easily when books with different conventions are being compared.

The above definition of **B** in a magnetized material as the total effective flux density due to all sources, is consistent with the use of **B** in the calculations of induced e.m.f. which culminated in Maxwell's equation for curl **E**:

$$\text{curl } \mathbf{E} = -\frac{d\mathbf{B}}{dt} \tag{7.25}$$

We should expect induced electric fields and induced e.m.f.s to depend upon the rate of change of total flux through the circuit, including the flux due to magnetization of any material which is present. Therefore the equations which are given in Chapter 7 are valid for media other than vacuum, provided **B** is interpreted as $\mu_0(\mathbf{H}+\mathbf{M})$.

Similarly, for Gauss's law in magnetism, of which equation **(8.10)** is the differential form, it is the total flux density from all sources, that must have zero divergence if there are to be no free magnetic poles. A simple argument shows that if **M** has a non-zero divergence, this leads to the presence of separated (not free) poles with density equal to $-\mu_0$ div **M**. This must be balanced by the pole density due to an equal and opposite value of μ_0 div **H**, so that the total pole density is:

$$0 = \mu_0 \text{ div } (\mathbf{H}+\mathbf{M}) = \text{div } \mathbf{B}$$

·2   Permeability

Our equations for free space need one modification before they can be conveniently applied to magnetizable media. Since we have broken the simple relationship between **B** and **H** by including the effects of the magnetization in one but not in the other, we must put:

$$\mathbf{B} = \mu\mu_0\mathbf{H}, \tag{8.18}$$

where μ is a new dimensionless constant which is called the permeability.
If this is substituted in:

$$\mathbf{B} = \mu_0(\mathbf{H}+\mathbf{M}) \tag{8.17}$$

with:

$$\mathbf{M} = \chi_m \mathbf{H} \tag{8.11}$$

the following relationship is obtained between the dimensionless constants μ and χ_m:

$$\mu = 1 + \chi_m \tag{8.19}$$

Thus in a diamagnetic substance, with χ_m small and negative, μ is slightly less than 1; in a non-magnetic material, χ_m is zero, and μ is precisely 1; and in a para-magnetic substance, χ_m is small and positive, making μ a little greater than 1. Since we expect induced e.m.f.s to depend on rate of change of total flux, we may allow for any magnetization by including a factor μ in all the equations for self-inductance and mutual inductance which we derived in Chapter 7. In the expression for internal self-inductance of a wire the factor μ will be the permeability of the metal, whereas in all the expressions for external self-inductance μ is the permeability of the surrounding medium. This really becomes important when part of the 'surrounding medium' is ferromagnetic; however, it may not be sufficient to treat this case by using a constant permeability. This will be discussed in the next section.

8·6   Ferromagnetism and hysteresis

8·6·1   *Hysteresis*

This chapter started with a reference to permanent magnets. The ability of these magnets to retain magnetization in the absence of a magnetizing field is one of the characteristics of the class of materials which are called ferromagnetic. It is part of a more general phenomenon known as hysteresis, in which the magnetization M, instead of being proportional to H, lags behind it, so that for any value of H, the magnetization depends on the process by which the value of H was reached.

Without going into the advanced theory of ferromagnetism, we may describe what is going on qualitatively as follows: within a crystalline region called a domain, alignment of atomic dipoles takes place cooperatively, so that all are parallel to each other and to one of a number of directions defined by the crystal structure. Under the influence of an external magnetic field, the magnetization of a domain may change to the direction which is most nearly parallel to the field; but such a change does not take place in the reversible manner characteristic of the alignment of paramagnetic atoms. It does not happen at all until the field is strong enough to overcome the cooperative tendency of the atomic moments to remain parallel to each other; but when one atomic moment changes direction, the whole domain goes with it. In a real specimen, many such domains may exist, each of volume 1 mm.^{3} or less, with their crystallographic directions un-correlated. Unmagnetized, such a specimen will have domains magnetized in different directions, so that the specimen has no net magnetic moment, but a moderately strong applied field H will cause each domain to take up the direction of magnetization which is most nearly parallel to the field, and the material is then said to be saturated. In such a material, M is often very much larger than $\mu_0 H$, so that M provides nearly the whole of B.

The process of reaching saturation in an increasing magnetic field involves the successive lining-up of the various domains, leading to a gradual, but not linear, increase in the overall magnetization of the specimen; this is shown by the dashed curve of figure 23. The rest of figure 23 shows the way in which B varies when H

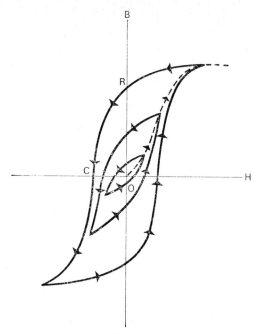

Figure 23.   Hysteresis curve.

is reduced to zero, then taken to large negative values, and finally increased through zero to large positive values. The complete curve thus traced out is called a hysteresis loop. The area of the loop is a measure of the energy per unit volume which is dissipated in taking the specimen through a complete cycle of magnetization. When the magnetic field is not taken to large enough values to cause saturation, a smaller hysteresis loop is obtained; two such loops are shown in figure 23.

The other important parameters of the hysteresis loop are the intercepts on the axes. OR, the value of B remaining when the applied magnetic field is reduced from maximum to zero, is called the remanence, or retentivity; OC, the value of negative H needed to reduce B to zero, is called the coercivity.

Materials for making permanent magnets

A material for making permanent magnets should have a large retentivity; but even more important, it must have a large coercivity, so that it can retain its magnetization in the presence of moderate demagnetizing fields, the most important of which is that resulting from its own magnetization. Such properties are found especially in cobalt steel, and in ferromagnetic alloys like Alnico (Al, Ni, Co, Cu, Fe) which are now widely used for making small permanent magnets.

Magnetostatics

3·6·3   Soft ferromagnetic materials

Another class of ferromagnetic material (sometimes called soft) has low coercivity and a narrow hysteresis loop, as shown in figure 24. The relatively easy magnetization and demagnetization is achieved by having domains whose boundaries are capable of movement, so that a reversible magnetization can take

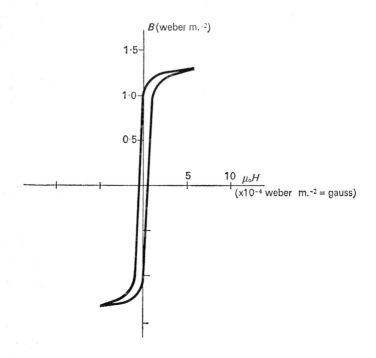

Figure 24.   Hysteresis curve of a soft ferromagnetic (silicon steel).

place by one domain growing at the expense of another. Since with these materials, a large B can be obtained with a small H, almost reversibly, we can describe their behaviour approximately by means of a permeability. In practice different values of permeability have to be used for different extreme values of H; a given material may have effective permeability of 1000 for small fields, the value dropping steadily as saturation is approached. It is substances of this nature which are important in the construction of transformers and self-inductances. With a core of silicon steel, or one of the special alloys called permalloy, a given arrangement of coils can have flux linkage, and hence inductance, a thousand times greater than the value obtained if it is wound on a non-magnetic former.

Soft ferromagnetic materials are useful also as cores for electromagnets: if we have a loop of iron of length L, with a gap of length l, as in figure 25, and wind

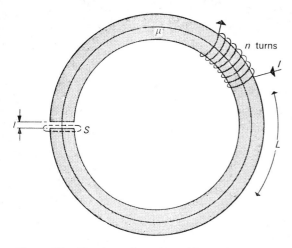

Figure 25.   Flux in an iron ring with an air-gap.

on it a coil of n turns, we have a simple electromagnet. When the coil is carrying a current I, we know from equation **(5.11)** that the line integral of magnetic field around the path shown dashed in figure 25 must be:

$$\oint \mathbf{H} \cdot \mathbf{ds} = nI$$

By applying Gauss's law to the closed surface S, it can be seen that the value of B in the gap must be the same as the value of B in the iron. Expressed in terms of this value of B, the magnetic field in the iron is $\dfrac{B}{\mu \mu_0}$, while the magnetic field in the gap is $\dfrac{B}{\mu_0}$ (assuming that the gap is effectively vacuum). The line integral will then be:

$$\frac{LB}{\mu \mu_0} + \frac{lB}{\mu_0} = \left(\frac{L}{\mu} + l\right) \frac{B}{\mu_0}$$

Putting this equal to nI, we get:

$$B = \frac{\mu_0 nI}{\left(\dfrac{L}{\mu} + l\right)}$$

This means that in the gap:

$$H = \frac{nI}{\left(\dfrac{L}{\mu} + l\right)} \tag{8.20}$$

This is usually very much larger than the magnetic field $\dfrac{nI}{2a}$ which would have been obtained if it had been possible to concentrate the n turns into an air-cored coil of radius a, around the place where the field was required.

Chapter 9
Further properties of the electric field

9·1·1   In Chapter 2, we considered some of the elementary properties of electric fields. We now return to electrostatics, to study in more detail the distribution of electric field around a charged object. The simplest charged object is a point

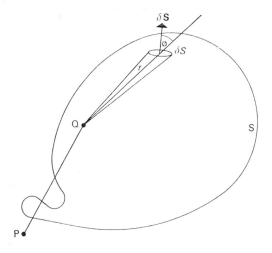

Figure 26.   Notation for Gauss's law.

charge Q, such as is shown in figure 26 with a closed surface S drawn around it. This surface is only a mathematical surface introduced for the sake of discussion; there need be no physical boundary at it.

The electric field due to Q is everywhere directed radially outwards from it, and its magnitude is given by equation (**2.4**) as:

$$E = \frac{Q}{4\pi\varepsilon_0 r^2} \tag{9.1}$$

In analogy to the magnetic flux density **B**, which in free space is equal to $\mu_0\mathbf{H}$, we now introduce the electric flux density, equal to $\varepsilon_0\mathbf{E}$ in free space. In the case under consideration, **D** also is directed radially outward from Q, and its magnitude is

$$D = \frac{Q}{4\pi r^2} \tag{9.2}$$

From the quantities on the right-hand side of equation (9.2), we see that electric flux density may be measured in coulombs per square metre. If there is a uniform flux density D across an area A, it can be said that there is, across A, an electric flux equal to the product DA. Since D is measured in coulombs per square metre, and A in square metres, the electric flux will be measured in units equivalent to the coulomb. The exact meaning of this will now be shown.

9·1·2 Taking a small area δS of our mathematical surface S, and describing it by a vector $\delta \mathbf{S}$ with direction along the normal to the area, and length defining its magnitude, we find that if the little area δS is inclined so that its normal makes an angle ϕ to the radius vector from Q, the flux across δS is equal to δS times the component of flux density along the normal to the area. This flux may be written as $D\delta S \cos \phi$, or alternatively as the scalar product $\mathbf{D} \cdot \delta \mathbf{S}$. Already some correspondence can be seen between this discussion and that of section 8·2, which dealt with Gauss's law in magnetism. Following the line of the magnetic case, we now calculate the total flux outward over the whole surface S; this is the sum of the scalar products $\mathbf{D} \cdot \delta \mathbf{S}$ for all the small areas δS which together make up the closed surface S. Substituting for D, we obtain for each scalar product:

$$\frac{Q\delta S \cos \phi}{4\pi r^2}$$

The quantity $\dfrac{\delta S \cos \phi}{4\pi r^2}$ is purely geometrical. $\delta S \cos \phi$ is the area cut off on a sphere of radius r with its centre at Q by the cone subtended at Q by the area δS; and $\dfrac{\delta S \cos \phi}{r^2}$ is the area cut off by this same cone on a sphere of unit radius, with centre at Q. This is a quantity which is given the special name of the solid angle subtended by δS at Q. The flux outward across δS is therefore equal to $\dfrac{Q}{4\pi}$ times the solid angle subtended by δS at Q. When all the scalar products are finally added to give the total flux outward over the surface S, we obtain $\dfrac{Q}{4\pi}$ times the total solid angle subtended at Q by the surface S. This total solid angle is the whole area of the sphere of unit radius, which is equal to 4π. The total flux outward over S is therefore numerically equal to Q, the factors 4π having cancelled. This holds whatever the shape of the closed surface S for even if it contains convolutions so that a given radius vector crosses it three times (as does QP in figure 26) instead of just once, some elements of flux simply leave, re-enter and leave again, thereby making the same contributions to the total as if they had left directly.

We may generalize also to the case of a surface containing several charges. Charges Q_1, Q_2, Q_3 ... Q_n have fields which can be superposed to give the total field, and cause total outward fluxes Q_1, Q_2, Q_3 ... Q_n which can be added to give the total outward flux, whether the charges are at the same or different points. Another charge Q_0, outside S, will give no total outward flux, since each element of flux leaving it which enters S will leave on the other side: entering and leaving, it will make equal and opposite contributions, which total zero.

It may therefore be stated, as a general conclusion which is known as Gauss's law (in electrostatics, to distinguish it from the magnetic case), that the total electric flux outward over a closed surface is, in all circumstances, numerically equal to the total electric charge inside. This may be written as an equation:

$$\int_s \mathbf{D}.\mathbf{ds} = \Sigma Q \tag{9.3}$$

where $\int_s$ means surface integral over a closed surface S, and Σ signifies the sum over all the charges inside S.

Gauss's law is not an accident: it arises from the fact that the inverse square of distance in the law of electrostatic force exactly cancels the factor r^2, which relates the areas cut off on concentric spheres by a cone with apex at the centre. As we have already pointed out when discussing the magnetic case, which also contains an inverse square law of force, such a law leads inevitably to a form of Gauss's law. The student may find it interesting to think about the gravitational analogue of Gauss's law.

9·2 Maxwell's equation for div D

When Gauss's law in magnetism had been established (in section 8·2), it was extended to a small volume and Maxwell's equation for div $\mathbf{B}$ was obtained (section 8·3). Similarly, now that Gauss's theorem in electrostatics has been established, it may be applied to a small volume by arguments precisely parallel with those of section 8·3. The argument for the electrostatic case is identical with that for the magnetic, up to equation (8.9); without repeating any of the argument, this can be rewritten in terms of electric flux instead of magnetic:

$$\text{Total electric flux} = \left(\frac{\partial Dx}{\partial x} + \frac{\partial Dy}{\partial y} + \frac{\partial Dz}{\partial z} \right) \delta x \, \delta y \, \delta z \tag{9.4}$$

The term in the bracket is div $\mathbf{D}$ and $\delta x \delta y \delta z$ is the volume V, so this can be replaced by the equivalent statement:

$$\text{Total electric flux} = V \text{ div } \mathbf{D} \tag{9.5}$$

Here the electrostatic case diverges from the magnetic, because the total electric flux leaving the surface of the parallelepiped is not necessarily zero, but is given by Gauss's law [equation (9.3)] as being numerically equal to th charge Q inside. So equation (9.5) becomes:

$$\text{div } \mathbf{D} = \frac{Q}{V} \tag{9.6}$$

$\frac{Q}{V}$ is the charge per unit volume inside the parallelepiped. If there are charge distributed uniformly over a volume larger than V, we may replace $\frac{Q}{V}$ by a charg density, or charge per unit volume, which is called ρ; hence:

$$\text{div } \mathbf{D} = \rho \tag{9.7}$$

This is the fourth Maxwell equation. It tells us that anywhere that $\mathbf{D}$ is found t have a non-zero divergence, there must be free charges distributed with a densit numerically equal to div $\mathbf{D}$. Together with the other three Maxwell equation [(5.15), (7.25) and (8.10)] it provides a complete description of the behaviour c electric and magnetic fields, from which the whole of electromagnetic theor can be deduced.

9·3 Effect of dielectric

In a material medium, an electric field can cause electric polarization, whic may be treated by methods closely analogous to those we developed for magneti zation. However, we shall not rely upon the analogy, because ordinary electri polarization is not a dynamic effect like diamagnetism, nor a process of alignmen like that which causes para- and ferromagnetism; it is a creation of atomi electric dipoles, by slight separation of the centres of gravity of the negative an positive charges. We define electric dipole moment as the distance of separatio of the equal and opposite charges multiplied by their magnitude. When a speci men contains many atomic electric dipoles, it has a total electric dipole momen equal to the vector sum of the atomic electric dipole moments. Thus we may sa that it has an electric polarization $\mathbf{P}$, equal to the electric dipole moment pe unit volume.

On a slab of dielectric, as shown in figure 27, polarization $\mathbf{P}$ gives surfac charges $\pm P$ per unit area. These create an electric field inside the material o magnitude $\frac{1}{\varepsilon_0} P$, in a direction opposite to that of the external field $\mathbf{E}_{\text{ext}}$ causin the polarization. (Note the failure of the magnetic analogy over the direction c this field). The same result is obtained, by less direct argument, for specimens c other shapes. The total electric field inside the material (averaged over loca variations) is therefore:

$$\mathbf{E}_{\text{int}} = \mathbf{E}_{\text{ext}} - \frac{1}{\varepsilon_0} \mathbf{P}$$

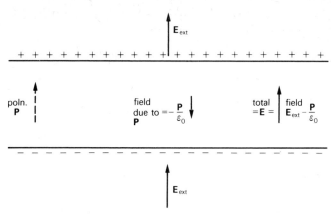

Figure 27. Electric field and polarisation in a dielectric.

To preserve the definition of E as the quantity whose line integral gives the work done in moving a test charge along a path, we must use E_{int} as the electric field E in the material.

Gauss's law shows that $\varepsilon_0 \, \text{div} \, E$ is the density of all charges in the material, both free and bound. The bound charges, which contribute to the polarization, have a density equal to $-\text{div} \, P$. Therefore, if we want to keep the flux density D as the quantity whose divergence is the density of free charges, we must put:

$$D = \varepsilon_0 E + P \qquad (9.8)$$

This makes D equal to $\varepsilon_0 E_{ext}$, the flux density that would have been produced by the same external charge-distribution, in the absence of polarization. It is also equal to ε_0 times the field in a flat cavity cut in the polarized dielectric, at right-angles to P.

When the polarization is caused by the presence of an electric field E in the material, it is usually proportional to it. The constant of proportionality, χ_e, is called the electric susceptibility, and is usually made dimensionless by including a factor ε_0, as follows:

$$P = \chi_e \varepsilon_0 E \qquad (9.9)$$

The corresponding magnetic equation (8.11) contained no μ_0, because M was given the dimensions of H, not of B, while P has been given the dimensions of D.

If we express the flux density D as:

$$D = \varepsilon \varepsilon_0 E \qquad (9.10)$$

where ε is a dimensionless constant known as the dielectric constant of the material, we get:

$$\varepsilon = 1 + \chi_e \qquad (9.11)$$

Equation (9.11) is closely analogous to equation (8.19), which relates magnetic permeability with magnetic susceptibility. There is, however, a difference in

Further properties of the electric field

magnitudes as well as in origin. Most solids and liquids have electric susceptibility between 1 and 10, with a few between 10 and 100. Dielectric constants over 2 are thus almost universal for solids and liquids, while gases at atmospheric pressure have dielectric constants around 1·001. The corresponding paramagnetic susceptibilities and permeabilities are much smaller. If we want to bring in ferromagnetic materials, with effective permeabilities of order 1000, we must compare them with the special class of materials which are by analogy called ferroelectric. The best known of these is barium titanate, with a dielectric constant of 1200.

In sections 9·1 and 9·2 we have discussed Gauss's law in electrostatics in its integral form, and also in the differential form which gives Maxwell's equation for div **D**. The results obtained in this discussion are applicable to systems containing polarizible materials as well as to free space, provided that **D** is defined by equation (9.8) to include the effect of polarization. If the polarization has a non-zero divergence, this leads to a volume density of polarization charge, which can either be cancelled by div $(\varepsilon_0 \mathbf{E})$, or be treated as a real charge density, reflected by the existence of a non-zero div **D**.

9·4 Electric field near a charged surface

Suppose an electric charge is spread over the surface of a metal in such a way that the charge per unit area has a uniform value σ. The system is taken to be in a steady state, i.e. no electric currents are flowing. Gauss's law is now applied to the mathematical closed surface S enclosing a small area δA of the real surface as shown in figure 28. Half of S is inside the metal. This means that there can be

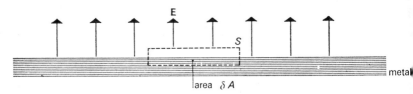

Figure 28.   Electric field near a charged conducting surface.

no electric flux over it, because a non-zero flux there would imply a non-zero electric field in the metal, which would cause a current to flow, thus contradicting the requirement that the system is in a steady state. Outside the metal, however, there can be an electric field **E**; at points close enough to the metal for the surface to look like an infinite plane, symmetry requires that **E** should be perpendicular to the surface. Provided the sides of S are perpendicular to the surface and short enough for this condition to apply, there will be no component of **E** across them, and hence no flux over them. So the outer surface of S which has area δA, a right-angles to an electric field of magnitude E, will provide the only non-zero contribution to the electric flux out of S. The magnitude of this is $\varepsilon \varepsilon_0 E \delta A$, where ε is the dielectric constant of the medium outside the metal. The charge inside S is $\sigma \delta A$, therefore by Gauss's law:

$$\varepsilon \varepsilon_0 E \delta A = \sigma \delta A$$

whence:

$$E = \frac{\sigma}{\varepsilon \varepsilon_0} \tag{9.12}$$

This is a useful general expression for the static electric field near to the surface of a charged conductor. It is valid for distances small compared with the radius of curvature, and the size, of the surface.

A curious fact follows from the above argument: there can be no static electric charge inside a conductor. In other words there can be no preponderance of charge of one sign extending over a distance large enough for the term conductor to be meaningful. Any electrostatic charge carried by a conductor must therefore be on its surface, where it has a possibility of creating the electric flux which is inevitably associated with it.

·5 Field due to a spherical distribution of charge

Let us consider a total charge Q, distributed with spherical symmetry. This means that the distribution has a centre, and that the distribution is the same with respect to all lines drawn outward from this centre. The distribution may be uniform over a certain volume, or it may be a single spherical shell, or it may be a series of concentric spherical layers, like the ideal onion. The radial distribution which distinguishes these cases is not specified, except by saying that all the charge is inside a radius R. At a point outside this radius R, it can be seen from the symmetry of the configuration, that the electric field can be in no direction except radially towards or away from the centre; also, at all points at radius r, greater than R, the electric field will have the same magnitude. So if Gauss's law is applied to the mathematical sphere of radius r, the total outward flux must be equal to Q. The area of the sphere is $4\pi r^2$, therefore the flux density over it is:

$$D = \frac{Q}{4\pi r^2}$$

and the magnitude of the electric field at it, if the medium is vacuum, is:

$$E = \frac{Q}{4\pi \varepsilon_0 r^2}$$

The vector electric field is obtained by including the unit vector $\dfrac{\mathbf{r}}{r}$, as follows:

$$\mathbf{E} = \frac{Q}{4\pi \varepsilon_0 r^2} \frac{\mathbf{r}}{r}$$

This equation is the same as equation (2.4) for the electric field due to a point charge Q. The exact correspondence shows that the electric field of any spherically symmetric distribution of charge, at points outside, is the same as it would be if the charge were concentrated at the centre. This is a very useful and important fact, which allowed the law of force between point charges to be inferred from measurements of the forces between charged spheres, and also allows a great simplification in the calculation of the electrostatic forces between particles or ions, which are spherical but not infinitesimally small.

A similar simplification in gravitational calculations may be obtained by treating spherical bodies as point masses.

9·6 Field inside a hollow charged conductor

Let us consider two conceivable configurations of electric field inside a hollow conductor carrying an electrostatic charge:

(i) A field which is directed from every point on the conductor towards a single point or region inside. If a mathematical surface is drawn round this region, we see that the field implies a net electric flux inwards over the surface, and therefore from Gauss's law an electric charge in the region towards which the field is directed. Therefore a field of this type cannot exist unless an isolated charge is suspended inside the conductor.

(ii) A field which is directed inwards from some parts of the conductor, but outwards towards it at other parts: with a field distribution of this type, a path could be found which started at a point on the conductor and ended at another and along which the line integral of E was non-zero. When the line integral of E was discussed in Chapter 2, it was called the potential difference between the end and the beginning of the path. A non-zero potential difference between two points on the conductor would cause a current to flow in it, changing the distribution of electric charge. So it follows that no static distribution of electric charge on a hollow conductor can give rise to this type of electric field inside.

Since all configurations of field inside must be in one or other of the two broad categories which we have shown to be impossible for a static distribution of charge on the conductor, with no other charged object inside, we are forced to conclude that there can be no electric field inside such a charged conductor. In other words, that the electric field is zero everywhere inside a charged conductor.

This fact provides the basis for several ideas, with different levels of importance. There is the technically important device of electrostatic screening, in which a metal cover ensures that apparatus is not disturbed by external electrostatic fields, nor indeed by electromagnetic radiation from outside. Of fundamental rather than practical significance is the fact that the absence of electric field inside a charged conductor (usually a metal sphere) provides a delicate test of the inverse square law. Experiments to detect any small electric field which may exist inside a charged metal sphere have been made by Cavendish, by Maxwell, and by Plimpton and Lawson. These have successively increased in accuracy, the last-mentioned having shown that the law of electrostatic force must involve a power of distance equal to 2, within 1 part in 10^9.

Chapter 10
Capacity

Definition and unit

10·1·1 *Capacity of an isolated sphere*

So far electric charges and the potential differences between pairs of points which may result from the presence of the electric charges have been discussed, but not much has been said about the potentials, or potential differences, of objects actually carrying electric charges. However, we are now in a position to say something meaningful about them, on the basis of section 9·5.

Suppose we have a conducting sphere of radius a, carrying a charge Q: section 9·5 shows that at points outside, this will have an electric field identical with that which would be observed if it were replaced by a point charge Q at its centre. If the fields are the same, the potential differences, which are simply line integrals of the field, must also be the same. Since the potential at a distance r from a point charge Q, with respect to a point at infinity, is known to be $\dfrac{Q}{4\pi\varepsilon_0 r}$ [from equation (2.6)], it follows that the potential at a distance r from the centre of our charged sphere is also $\dfrac{Q}{4\pi\varepsilon_0 r}$. This must hold even when r is only infinitesimally larger than the radius a, and as the electric field increases smoothly on moving towards the surface of the conducting sphere, without any sudden jump as we reach the sphere itself, the potential must have the same value *on* the sphere as it has at a point infinitesimally close to it. This is $\dfrac{Q}{4\pi\varepsilon_0 a}$, and it represents the work per unit charge which would have to be done in bringing up an additional small charge from infinity to the sphere.

For the present we are interested in the fact that the potential of the sphere is proportional to the charge on it. Conversely, if we cause the sphere to be at a given potential, it will have to carry a charge proportional to this potential. The constant of proportionality, charge/potential, is called the capacity. For the isolated sphere, it has been shown that this is equal to:

$$C = 4\pi\varepsilon_0 a \qquad\qquad (10.1)$$

In section 6·6, we have mentioned that ε_0 has the value $\dfrac{1}{36\pi \times 10^9}$ farad per metre; we now see that the farad is a unit of capacity. In fact, an isolated spherical conductor of radius r (in metres) has capacity in farads given by:

$$C = \frac{r}{9 \times 10^9}$$

A sphere of radius 1 cm. therefore has a capacity $\dfrac{1}{9 \times 10^{11}}$ farads, which is not very different from a micro-microfarad or picofarad (10^{-12} farad). The centimetre (i.e. capacity of a sphere of radius 1 cm.) is used as a unit of capacity in the c.g.s. electrostatic system of units; transition to farads may be made with the factor 10^{12}, with or without the further factor 0.9 according to the precision required.

10·1·2 *Condensers*

It can be seen that, at least for isolated spheres, the farad is not a very practical unit, as it is the capacity of a sphere of radius 9×10^9 metres. This is 30 per cent larger than the sun, which even if not convenient for experimentation is after all perhaps not too bad an approximation to an isolated conducting sphere.

In fact we are not often concerned with the capacities of isolated objects; more important is the situation where two neighbouring conductors carry equal and opposite charges, by virtue of a potential difference between them. Such a system is called a condenser, or capacitor, and is said to have a capacity $\dfrac{Q}{V}$, where the conductors carry charges $\pm Q$, and V is the potential difference between them. Equal and opposite charges are specified because a flux Q can start on a charge $+Q$ and end on a charge $-Q$, making a self-contained distribution of electric field. However, in practice the charges $\pm Q$ are not necessarily the total charges carried by the conductors; they are the charges associated with the operation of these two conductors as a condenser of capacity C. Either conductor may be capable of carrying additional charge through acting as a condenser of capacity C' with some third conductor such as the earth. For example, in a pair of concentric spheres of capacity C, all the flux leaving the inner sphere must reach the outer; so the inner sphere would carry only a charge $+Q$ equal to C times the potential difference V between the spheres. The outer sphere would carry a charge $-Q$ on its inner surface, where the flux from the inner sphere ends, through its operation as part of the condenser of capacity C; but it might also carry, on its outer surface, a charge Q' equal to $C'V'$, where C' is the capacity of the condenser made up of the outer sphere and the earth and V' is the potential of the outer sphere with respect to the earth.

As will be seen in the following sections, capacities for pairs of conductors arranged as condensers tend to be greater than they are for isolated conductors of the same dimensions. Therefore the farad is a marginally less unpractical unit for condensers than it is for isolated objects.

·2 General method of calculating capacities

·1 *Principles of the method*

For two conductors forming a condenser of given geometry, the capacity can be found either by postulating a given charge and calculating the potential difference, or by postulating a potential difference and calculating the charge. Either of these methods allows the capacity to be calculated as a ratio, but the former is usually the easier. The general method is as follows:

(i) Choose one of the conductors and put a charge Q on it (if the conductors are one inside the other, the charge must be placed on the inner one).

(ii) Use Gauss's law to state that the flux leaving this conductor is equal to Q.

(iii) Study the symmetry of the system in order to determine how this flux will be distributed, and calculate the electric field at a general point between the conductors.

(iv) Take the line integral, from one conductor to the other, of this expression for the electric field; the result will be the potential difference V between the conductors.

(v) Divide Q by V to obtain the capacity C.

In the following three sections we shall demonstrate this method by working out three examples. In cases which do not have enough symmetry to allow step (iii), approximations may help, or the problem may be soluble as a superposition of two symmetrical cases. In really difficult cases, experimental measurement may be easier and more realistic than calculation. Such measurement may be direct measurement of capacity (see Chapter 16), or it may be the measurement of the electrical resistance of a model placed in an electric trough. The correlation between these two measurements will now be shown.

2 *The electrolytic trough*

Suppose that we make a full-scale model of our condenser and fill the space between the electrodes with a medium of high, but not infinite, electrical resistivity ρ. If we apply the same potential difference V to the model as to the original, the distributions of electric field and equipotential surfaces will be the same. But in the model, small currents will flow. If the electric field is $\mathbf{E}$, the current per unit area, $\mathbf{j}$, will be equal to $\dfrac{1}{\rho}\mathbf{E}$. The total current, I, flowing between the electrodes of the model can be calculated by taking the surface integral $\int \mathbf{j} \cdot \mathbf{dS}$ over a closed surface separating one electrode from the other. This gives:

$$I = \int \mathbf{j} \cdot \mathbf{dS} = \frac{1}{\rho}\int \mathbf{E} \cdot \mathbf{dS}$$

If we now replace $\mathbf{E}$ by $\dfrac{1}{\varepsilon\varepsilon_0}\mathbf{D}$, where $\mathbf{D}$ is the flux density at the corresponding point in the original condenser, and ε the dielectric constant of the material between the plates of the original condenser, we find that:

$$I = \frac{1}{\rho\varepsilon\varepsilon_0} \int D.dS$$

Now $\int D.dS$ is the total electric flux from one electrode to the other, in the condenser, and Gauss's law tells us that this must be equal to the magnitude Q of the charge on each. Therefore for a given potential difference V, the current I in the model is related to the charge Q on the original condenser by:

$$I = \frac{1}{\rho\varepsilon\varepsilon_0} Q$$

The resistance R of the model is thus related to the capacity C of the original by:

$$R = \frac{V}{I} = \rho\varepsilon\varepsilon_0 \frac{V}{Q} = \frac{\rho\varepsilon\varepsilon_0}{C}$$

and conversely:

$$C = \frac{\rho\varepsilon\varepsilon_0}{R} \tag{10.2}$$

Therefore the capacity of the condenser is equal to $\rho\varepsilon\varepsilon_0$ divided by the resistance of an identical system with the material of dielectric constant ε replaced by a conducting material of resistivity ρ. This fact, coupled with the possibility of scaling down the dimensions, allows the capacities of complicated systems to be estimated by means of models immersed in a liquid of high resistivity. The liquid chosen is a dilute electrolyte. The method is particularly useful for finding the capacities of objects such as valves which are not designed as condensers.

10·3 Concentric spheres

The general method for calculating capacities, outlined in the preceding section, can be applied to the case of two concentric metal spheres (see figure 29) by putting a charge Q on the inner sphere; the electric flux leaving it will then be Q. This is the most symmetrical case of all, for if we draw a mathematical sphere of radius r between the metal spheres, the flux Q must be spread uniformly over the area of the sphere, which is $4\pi r^2$. The flux density is therefore $\frac{Q}{4\pi r^2}$, just as if the charge were concentrated at the centre. If the space is filled with a material of dielectric constant ε, the electric field will be:

$$E = \frac{Q}{4\pi\varepsilon\varepsilon_0 r^2}$$

radially outwards.

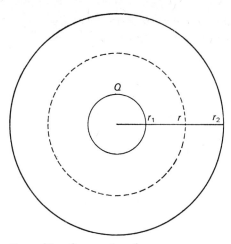

Figure 29. Concentric spheres.

To obtain the potential difference between the metal spheres, to which we assign radii r_1 and r_2, we have to take the line integral of $\mathbf{E}$ between them. However, since the field is radial, the line integral of $\mathbf{E}$ is just the ordinary integral of E with respect to r, and the potential difference is:

$$V = \int_{r_1}^{r_2} E \, dr$$

$$= \frac{Q}{4\pi\varepsilon\varepsilon_0} \int_{r_1}^{r_2} \frac{dr}{r^2}$$

$$= \frac{Q}{4\pi\varepsilon\varepsilon_0} \left(\frac{1}{r_1} - \frac{1}{r_2} \right)$$

The last step is to take the ratio $\dfrac{Q}{V}$, obtaining:

$$C = \frac{Q}{V} = \frac{4\pi\varepsilon\varepsilon_0}{\left(\dfrac{1}{r_1} - \dfrac{1}{r_2} \right)}$$

$$\therefore C = \frac{4\pi\varepsilon\varepsilon_0 r_1 r_2}{r_2 - r_1} \tag{10.3}$$

10·4   Coaxial cylinders

The same general method may be used for calculating the capacity of a condenser consisting of two coaxial cylindrical conductors. These might be in the form of a wire suspended along the axis of a round metal tube, or one tube held inside another by insulating annular spacers.

Initially, the cylinders are assumed to be long, and a charge Q per unit length is placed on the inner one; we then confine our attention to a piece near the middle, of unit length (see figure 30). The flux leaving this piece will be equal to

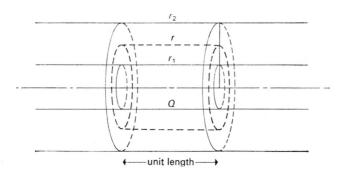

Figure 30.   Coaxial cylinders.

the charge on it, namely Q, and if the ends of the cylinder are so far away that they have no effect on the distribution of flux, this will be a uniform cylindrically symmetric distribution, with the flux in planes perpendicular to the axis, having no component parallel to the axis. So if we draw a mathematical cylinder of radius r, between the two cylindrical conductors and coaxial with them, the flux Q will be distributed uniformly over unit length of its surface, which has area $2\pi r$. The electric field at points distant r from the axis therefore has magnitude:

$$E = \frac{Q}{2\pi\varepsilon_0 r}$$

This field is directed radially outwards from the axis, so the line integral of $\mathbf{E}$ is again the ordinary integral of E with respect to r. Therefore the potential difference between the conducting cylinders, of radii r_1 and r_2, is:

$$V = \int_{r_1}^{r_2} E \, dr$$

$$= \int_{r_1}^{r_2} \frac{Q}{2\pi\varepsilon_0 r} \, dr$$

$$= \frac{Q}{2\pi\varepsilon_0} (\log_e r_2 - \log_e r_1)$$

$$= \frac{Q}{2\pi\varepsilon_0} \log_e \left(\frac{r_2}{r_1}\right)$$

The capacity of this unit length of the system is therefore:

$$C = \frac{Q}{V} = \frac{2\pi\varepsilon\varepsilon_0}{\log_e\left(\dfrac{r_2}{r_1}\right)}$$

If the cylinders have total length l, and we neglect the end corrections resulting from the spread of the field beyond the physical ends, the total capacity is:

$$C = \frac{2\pi\varepsilon\varepsilon_0 l}{\log_e\left(\dfrac{r_2}{r_1}\right)} \tag{10.4}$$

Since ε_0 is measured in farads per metre and l in metres, and ε and $\log_e\left(\dfrac{r_2}{r_1}\right)$ are dimensionless quantities, this gives C in farads, as required. For numerical calculations we may substitute the value $\dfrac{1}{36\pi \times 10^9}$ farads per metre for ε_0, obtaining

$$C = \frac{\varepsilon l}{18 \times 10^9 \log_e\left(\dfrac{r_2}{r_1}\right)} \text{ farad}$$

$$= \frac{1000\,\varepsilon l}{18 \times 2\cdot306 \log_{10}\left(\dfrac{r_2}{r_1}\right)} \text{ picofarad}$$

Here ε is the dielectric constant of the material between the cylinders; the material outside does not matter. By a numerical coincidence the dielectric constant of polyethylene is 2·3, so for a coaxial cable with polyethylene insulator, ε cancels with the factor for converting the logarithm from base 10 to base e. In fact this forms the commonest system of coaxial cylindrical conductors.

0·5   Parallel plates

In the case of two parallel plates of area A, separated by a distance d which is much less than their lateral dimensions, as the standard first step towards

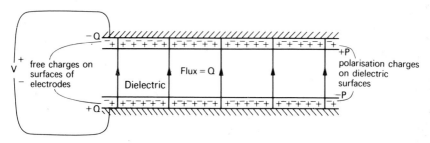

Figure 31.   Parallel-plate condenser. For edge effects see figure 1.

calculating the capacity, one of the plates is given a charge Q. We assume that this charge is all stored in the mutual capacity of the two plates, so that the flux from it ends on the other plate. For this to hold, the charge must be all on the inner surface (see figure 31); so if edge effects are neglected, it will be distributed uniformly at a density of $\dfrac{Q}{A}$ per unit area.

The field near to the surface may now be obtained either by using Gauss's law, or from equation (9.12) which has already been obtained by the use of this law. If the plates are close enough to each other for us to be able to neglect the effects of the edges, the flux which leaves the surface at a uniform density of $\dfrac{Q}{A}$ per unit area will go straight across to the other plate. The field will then have the same magnitude $\dfrac{Q}{\varepsilon\varepsilon_0 A}$ everywhere in the space between the plates, and the potential difference will be the simplest of all line integrals: the magnitude of the uniform electric field multiplied by the distance between the plates. This gives:

$$V = \frac{Qd}{\varepsilon\varepsilon_0 A}$$

and the capacity is given by the ratio of Q to V as:

$$C = \frac{\varepsilon\varepsilon_0 A}{d} \tag{10.5}$$

Numerically, this gives:

$$C = \frac{\varepsilon A}{36\pi d} \text{ millimicrofarad} \tag{10.6}$$

with A expressed in square metres and d in metres.

The effect of the solid dielectric shown in figure 31 is described by the dielectric constant ε in equations (10.5) and (10.6). An equivalent description in terms of surface charges is given in section 9·3.

10·6 Condensers in practice

10·6·1 *Effect of dielectric*

In practice, condensers are nearly always made with a solid material between the conductors; such a material serves three purposes. First, as an insulator, it serves to prevent the passage of current, even at voltages which would be large enough to cause a spark across an equivalent thickness of air; next as a dielectric with dielectric constant greater than 1, its polarization can serve to increase the capacity obtained with a given geometry; and thirdly, it can be valuable as a spacer, holding the conductors at a fixed uniform distance from each other.

Commonly used dielectric materials are impregnated paper, mica, polyesters and special ceramics; the conductors may be either sheets of metal foil, or thin metallic films deposited on the surfaces of the dielectric material. The important characteristics of a condenser are its capacity and the voltage which can be applied across it without risk of breakdown; the latter is determined by the thickness of the dielectric material and by the maximum electric field which it can withstand. The last-mentioned quantity is sometimes known as the 'dielectric strength' of the material, and is dependent upon the possibility of partial ionization of what is otherwise an insulator, in an intense electric field.

6·2   *Construction of real condensers*

The capacity, on the other hand, involves the area as well as the thickness of the dielectric material. Insofar as the arrangement behaves like a pair of parallel plates, the capacity is given by equation **(10.6)**.

Small capacities (picofarads) may be obtained with a few square millimetres of thin metal-coated dielectric, acting as a parallel-plate condenser. Larger capacities (up to a few microfarads) are achieved with larger areas of flexible metal-coated dielectric (especially paper), rolled up into a cylindrical shape. Condensers of this type become very clumsy for large capacities (hundreds or thousands of microfarads), so whenever possible these are obtained with electrolytic condensers, in which the dielectric material is a very thin layer of oxide, maintained by electrolytic action between a metal electrode and a conducting paste. The thickness of this layer can be very small and its area large, especially if the surface is artificially roughened; so large capacities can be obtained in condensers which are physically small. The chief limitation is that the voltage must be applied in a specified direction, since accidental connection to an alternating potential difference, or to a steady one in the wrong direction, can reverse the electrolytic process and destroy the condenser.

6·3   *Condensers in circuit*

When a condenser is treated as an element in a circuit, we ignore the details of its construction, and represent it by the symbol shown in figure 32(a). When

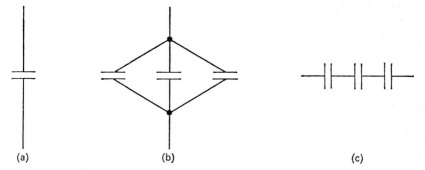

(a)   (b)   (c)

Figure 32.   Condensers (a) alone, (b) in parallel, (c) in series.

there are a number of condensers connected in parallel, as in figure 32(b), the charges Q_1, Q_2, Q_3 on them add to give the total charge Q. The total capacity is therefore the sum of the individual capacities:

$$C = \frac{Q}{V} = \frac{Q_1}{V} + \frac{Q_2}{V} + \frac{Q_3}{V} = C_1 + C_2 + C_3$$

But when several condensers are connected in series, as in figure 32(c), the charge Q on each is the same, and potential differences V_1, V_2, V_3 must be added to give the total potential difference V. The capacity C of the system is therefore given by:

$$\frac{1}{C} = \frac{V}{Q} = \frac{V_1}{Q} + \frac{V_2}{Q} + \frac{V_3}{Q} = \frac{1}{C_1} + \frac{1}{C_2} + \frac{1}{C_3}$$

10·7 Stored energy

When an extra charge δQ is transferred from one plate to the other of a condenser of capacity C, which is already charged to a potential difference V, an energy $V\delta Q$ must be provided by the agency transferring the charge. The total energy needed, to build up a charge Q from zero, must therefore be:

$$\int_0^Q V dQ = \int_0^Q \frac{Q}{C} dQ = \frac{Q^2}{2C}$$

This energy, which may alternatively be written $\frac{1}{2}QV$ or $\frac{1}{2}V^2C$, must be stored in the condenser, available for release during the discharge of the condenser.

Now suppose the condenser is of the parallel plate type with C equal to $\frac{\varepsilon\varepsilon_0 A}{d}$; there is inside it a uniform electric field E equal to $\frac{V}{d}$. The energy stored may be expressed as:

$$\tfrac{1}{2}V^2C = \tfrac{1}{2}E^2d^2 \frac{\varepsilon\varepsilon_0 A}{d} = \tfrac{1}{2}\varepsilon\varepsilon_0 E^2 Ad$$

Now Ad is the volume over which the uniform field E extends; we may therefore consider that the stored energy is distributed uniformly, at a density $\frac{1}{2}\varepsilon\varepsilon_0 E^2$ per unit volume. More general arguments lead to the conclusion that wherever there is an electric field E, there is a density of stored energy which may be expressed either as $\frac{1}{2}\varepsilon\varepsilon_0 E^2$ or as $\frac{1}{2}DE$. In free space this is $\frac{1}{2}\varepsilon_0 E^2$. We may at this point call to mind the very similar expression $\frac{1}{2}\mu_0 H^2$ [equation (7.6)] for the density of stored energy in a magnetic field.

Part Two
Circuits and currents

Chapter 11
Varying currents

1·1 The applicability of Kirchhoff's laws to varying currents

1·1 Application of Kirchhoff's first law

In Chapter 3, Kirchhoff's two laws were introduced as consequences of applying the principles of conservation of charge and energy to a circuit carrying a steady current. Now we must examine the implications of these principles for circuits in which the current is varying, and at the same time extend our discussion to circuits containing condensers and inductances. In doing this, the possibility of equal and opposite electrostatic charges building up on the plates of a condenser must be taken into account, but no other possible accumulations of charge on parts of the circuit will be considered. In other words, we stick to the idea of current flowing *in a circuit*; it may flow through a resistance or a self-inductance, in the sense that the same number of electrons leave one end of the component in a given time as enter it at the other end in the same time, because the wire is continuous and we have agreed to neglect local accumulations of charge. A current may flow through a condenser in the above sense, but in this case the reason is that the charges accumulating on the plates are equal and opposite (see section 10·1), and we have agreed to neglect local accumulations of charge which are not balanced in this way. In fact for a well-made condenser, the approximation made in neglecting them is no worse than for a component like a resistor.

We shall therefore consider Kirchhoff's first law as applying to the instantaneous values of currents at any moment, even if these are varying. In a simple circuit, this means that the current all the way round the circuit has the same value at any given moment; and in a circuit with side branches, it simply means that the algebraic sum of the instantaneous currents reaching a junction is zero at all times.

1·2 Application of Kirchhoff's second law

The principle of conservation of energy leads to a more general form of Kirchhoff's second law which looks very different from equation (3.4). In the general circuit there may be a source of e.m.f. supplying energy to the electrons which form the current. Some energy may be stored in the magnetic field of any self-inductance in the circuit, the amount stored increasing as the current increases. This extra energy is drawn from the current, through the induced e.m.f. set up in the inductance. However, when the current is decreasing, the induced e.m.f. is in the other direction, tending to maintain the current, and feeding energy back into the circuit at the expense of a reduction in the amount stored. Additional energy may be stored in any capacitative components, the amount changing according to the amount of charge stored on their plates; as in the case of the inductance, any increase in the amount stored is provided by the flow of the current against the potential difference across the capacity, and any decrease is fed back into the circuit by the flow of a current in the direction of this potential difference. In the resistive components, however, energy is dissipated and not stored. Therefore, for the whole circuit, conservation of energy requires that the rate of supply of energy from the source must be equal to the rate of increase of energy stored in the capacitative and inductive parts of the circuit, plus the rate of dissipation of energy as heat in the resistance of the various components. This is necessary because the electrons of the current form a negligible reservoir of energy, apart from capacitative and inductive effects; they can act as the agent for transporting energy from one part of the circuit to another, but what they are transporting is included in the total of stored energy while it is in transit.

11·1·3   Closed circuits within networks

This simple balance of rate of supply of energy against total rate of storage and dissipation of energy applies only to a complete network. Within a closed circuit which forms part of a larger network (e.g. figure 33), there need be no such

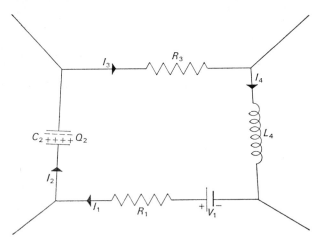

Figure 33.   Closed circuit which is part of a larger network

balance, since energy may be flowing into the circuit from the rest of the network or vice versa. But we can say something about the closed circuit: an individual electron going round this closed circuit must have total work done on it equal to zero; it may carry energy from source to store, or from store to be dissipated as heat, but it can itself suffer no net loss or gain of energy. Therefore the e.m.f. of the source must be balanced by the sum of all potential differences due to:

(a) current flowing through resistances
(b) accumulation of charge in condensers
and (c) induced e.m.f. due to changing currents in inductances.

Algebraically this requires, in analogy to equation (3.4),

$$V = IR + \frac{Q}{C} + L\frac{dI}{dt} \qquad \text{(11.1)}$$

If we keep the same sign convention as equation (3.4), this gives for the circuit of figure 33:

$$V_1 = I_1 R_1 + I_3 R_3 + \frac{Q_2}{C_2} + L_4\frac{dI_4}{dt}$$

I_2 itself does not matter, but Q_2 may be expressed as the time-integral of I_2, since it just represents the accumulated charge carried by I_2.

·2 Step-voltage applied to a self-inductance

·1 As a first exercise in the study of varying currents, we consider what happens when the switch is closed in the circuit of figure 34(a). This may represent the con-

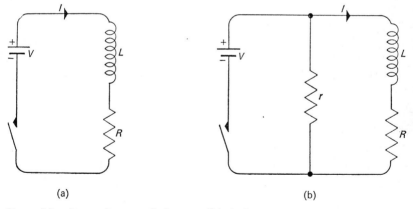

(a)   (b)

Figure 34.   Step-voltage applied to a self-inductance
(a) without parallel resistance
(b) with parallel resistance.

nection of a battery to something like an electromagnet, which has both self-inductance and resistance. If the switch is closed at time $t = 0$, the current at later times must satisfy the appropriate form of equation (11.1), which for this circuit is:

$$V = IR + L\frac{dI}{dt}$$

Since V, R and L are constants, this equation can be solved as follows:

$$\frac{V}{R} - I = \frac{L}{R}\frac{dI}{dt}$$

$$\therefore \int \frac{R}{L}dt = \int \frac{dI}{\left(\frac{V}{R} - I\right)}$$

$$\therefore \frac{R}{L}t = \text{constant} - \log_e\left(\frac{V}{R} - I\right)$$

$$\therefore \frac{V}{R} - I = e^{-\frac{R}{L}t} \times \text{another constant}$$

$$\therefore I = \frac{V}{R} - \text{constant} \times e^{-\frac{R}{L}t}$$

If I is to be zero at time $t = 0$, when the factor $e^{-\frac{R}{L}t}$ is unity, the constant must be $\frac{V}{R}$, and so:

$$I = \frac{V}{R}\left(1 - e^{-\frac{R}{L}t}\right)$$

$\frac{V}{R}$ is the value which the current will reach after an infinitely long time, and we may call it I_{max}; with this notation the current at a general time t is:

$$I = I_{max}\left(1 - e^{-\frac{R}{L}t}\right)$$

which shows an exponential approach toward its final value.

11·2·2   Now what happens if the switch is opened again? If L is the inductance of a large electromagnet the answer is probably disaster, for as the switch opens, the current starts to decrease rapidly and this sets up, in the self-inductance, a very large induced e.m.f. which tries to maintain the current. It will do this by causing a spark to jump, either across the opening switch, or if this is opening too fast across the insulation of the electromagnet itself. Whichever happens, if an attempt is made to switch off a current of several amperes in an inductance of several henries, over a period of the order of a millisecond, an induced e.m.f. of tens of thousands of volts will be set up and something is likely to be damaged.

For this reason electromagnets are not fed from sources which can be switched off directly; one uses a transformer-rectifier set in which the output current can continue to flow after the input has been disconnected, or else a rotating generator, the output of which can be controlled gently by adjustment of the current in the field windings. Alternatively, as a safeguard in experimental work, a resistance r may be connected permanently in parallel with the magnet [see figure 34(b)]. Power will be wasted in r while the switch is closed, but if the switch is opened there will still be a return path for the continued flow of current through the self-inductance. In these circumstances, the current I will decrease in such a way that the negative quantity $L\dfrac{dI}{dt}$ always exactly balances the positive potential difference $(r+R)I$; this leads to an equation which may be solved for I as follows:

$$(r+R)I+L\frac{dI}{dt} = 0$$

$$\therefore \quad -\frac{(r+R)}{L}\int dt = \int \frac{dI}{I}$$

$$\therefore \quad \log_e I = -\left(\frac{r+R}{L}\right)t + \text{constant of integration}$$

$$\therefore \quad I = \text{constant } e^{-\left(\frac{r+R}{L}\right)t}$$

If we now use $t = 0$ for the moment of opening the switch, the constant must be the current I_0 which was flowing at this moment, and we have an exponential decay of current according to the relation:

$$I = I_0 e^{-\left(\frac{r+R}{L}\right)t}$$

1·2·3 It is interesting to calculate the potential difference across the terminals of r: while the current is decreasing, this is:

$$V' = -Ir = -I_0 r\, e^{-\left(\frac{r+R}{L}\right)t}$$

The negative sign represents the fact that current is flowing through r in the sense opposite to that when the switch is closed. If the current had been flowing for a long time before the switch was opened, making I_0 equal to the limiting value I_{max} $\left(\text{which was earlier put to equal } \dfrac{V}{R}\right)$ we may substitute to get:

$$V' = -V\frac{r}{R} e^{-\left(\frac{r+R}{L}\right)t}$$

This means that if we made $r = R$, so that with the switch closed and the steady state reached, half the current supplied by the source would be wasted through r, opening the switch would cause the potential difference across r to jump from V to $V' = -V$, after which it would decrease exponentially to zero. But if r is made equal to $100R$, to keep the steady loss of power equal to only 1 per cent, then opening the switch causes the potential difference across r to jump from V to $-100V$. In practice, therefore, the parallel resistance r must be capable of standing the source voltage steadily, and something like ten times this for the duration of the exponential decay.

If r is the resistance of a lamp which is inserted for the extra purpose of indicating that the magnet is on, one may choose to have either a normally dim indicator lamp flashing into brightness if the circuit is broken; or else a normally bright indicator lamp burning out spectacularly when the circuit is broken.

Of course, a rectifier in parallel with the magnet affords a useful protection against such troubles. This is connected in the direction that offers high resistance to the source; but if ever the main circuit is broken, the current flowing in the magnet finds in the rectifier a return path of low resistance.

11·3 Time-constant of an inductive circuit

In the preceding discussion of exponential increase or decrease of current in a circuit containing inductance L and resistance R (or $r + R$), we had factors of the form $e^{-\frac{R}{L}t}$, which can be written in the form $e^{-\frac{t}{T}}$. This T is the time necessary for the difference between instantaneous current and final current to be reduced by a factor e (approximately 2·718); it is called the time-constant of the circuit and is equal to $\dfrac{L}{\text{total resistance}}$. If L is the inductance in henries and R the total resistance in ohms, $T = \dfrac{L}{R}$ is the time-constant in seconds. An air-cored coil with a self-inductance of 1 microhenry in a circuit of total resistance 1000 ohms, would therefore have a time-constant of 1 nanosecond ($= 10^{-9}$ second).

11·4 Charging and discharging of a condenser

Another simple example of a varying current is that involved in charging or discharging a condenser. These operations differ from those of connecting or disconnecting a battery from an inductive circuit, in that the final current through a condenser is zero, whether it is connected to a battery or not. Working from equation (11.1), or from the simple balance of the potential differences across the resistance and the capacity against the applied e.m.f., for the circuit of figure 35, with the switch closed in the left-hand position, we obtain:

$$V = IR + \frac{Q}{C}$$

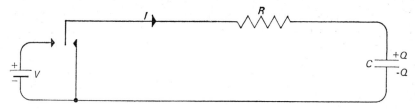

Figure 35.   Charging and discharging of a condenser.

This can be solved for Q by putting:

$$I = \frac{dQ}{dt},$$

or for I by putting:

$$Q = \int I dt$$

In practice the second method reduces to the first and either gives:

$$V = R\frac{dQ}{dt} + \frac{Q}{C}$$

$$\therefore VC - Q = RC\frac{dQ}{dt}$$

$$\therefore \int \frac{dt}{CR} = \int \frac{dQ}{VC - Q}$$

$$\therefore \frac{t}{CR} = -\log_e\ VC - Q) + \text{constant}$$

$$\therefore VC - Q = \text{constant } e^{-\frac{t}{CR}}$$

$$\therefore Q = VC - \text{constant } e^{-\frac{t}{CR}}$$

$$= VC\left(1 - e^{-\frac{t}{CR}}\right)$$

if $Q = 0$ when the switch is closed at time $t = 0$.

Thus Q, the charge stored on the condenser, approaches its final value VC exponentially.

Meanwhile, the current has a value given by:

$$I = \frac{dQ}{dt} = VC\frac{d}{dt}\left(1 - e^{-\frac{t}{CR}}\right)$$

3   **Varying currents**

$$= VC\left(\frac{1}{CR} e^{-\frac{t}{CR}}\right)$$

$$= \frac{V}{R} e^{-\frac{t}{CR}}$$

$$= I_0 e^{-\frac{t}{CR}}$$

where $I_0 = \dfrac{V}{R}$ is the initial current, equal to that which would flow indefinitely if the condenser were short-circuited. But having started with this value, the current is reduced exponentially as charge accumulates on the condenser, and builds up a potential difference opposing V.

The condenser can now be discharged by simply throwing the switch so that it is closed in the right-hand position; the discharge will then follow the equation that was used for the charging process, but with V put equal to zero:

$$IR + \frac{Q}{C} = 0,$$

which leads to

$$Q = Q_0 e^{-\frac{t}{CR}}$$

with Q_0 representing the initial charge. If discharge started after the condenser had been fully charged to a potential difference V, Q_0 is equal to EC. The discharging current is therefore:

$$I = \frac{dQ}{dt} = -\frac{Q_0}{CR} e^{-\frac{t}{CR}}$$

$$= -\frac{V}{R} e^{-\frac{t}{CR}}$$

Like the charging current, this starts at the same value as would be maintained steadily through a resistance R connected directly to a source of e.m.f. V. Thereafter it decreases exponentially to zero.

11·5   Time-constant of a capacitive circuit

Here again, we may express the time-variations of the exponentially-varying currents in terms of a time-constant T; this is the time required for a charge or a current to grow nearer to its final value by a factor $\dfrac{1}{e}$. The final value is never quite reached, but every interval T that elapses allows the departure from the final value to be reduced by a factor e. From the form of the exponential factors

$e^{-\frac{t}{CR}}$, we see that the time-constant for a capacitive circuit is CR.

With C expressed in farads, and R in ohms, CR would give T in seconds; but it is of more practical use to express C in microfarads and R in megohms, thus obtaining T in seconds, or C in picofarads and R in megohms to give the time-constant in microseconds. Such time-constants are of vital importance in the transmission of pulses through electronic circuits.

Chapter 12
Alternating current

12·1 Generation of alternating e.m.f.

In deriving the general expression (7.1) for the induced e.m.f. due to a changing flux through a circuit, we considered a coil of area A rotating with constant angular velocity $\dfrac{d\theta}{dt}$ in a uniform magnetic flux density B. The induced e.m.f. in this case was:

$$V = AB \sin \theta \, \frac{d\theta}{dt}$$

In Chapter 7, this was simply reduced to $-\dfrac{dN}{dt}$, the rate of change of flux at a particular moment. But now the variation of this e.m.f. with time will be considered.

If θ is replaced by ωt, where ω is the angular velocity $\dfrac{d\theta}{dt}$, the induced e.m.f. will be proportional to $\sin \omega t$. This means that it is varying sinusoidally, like the displacement in a simple harmonic motion. Such a sinusoidally-varying e.m.f. is known as an alternating e.m.f. If it is used to force a current through a resistance R, the current is equal to $\dfrac{V}{R}$ at each moment; it changes in magnitude and in direction, according to the same factor $\sin \omega t$ which is contained in V, and is known as an alternating current.

If an alternating current I is described by:

$$I = I_0 \sin \omega t,$$

we see that I_0 is the peak magnitude of the current, and ω is the angular frequency, equal to $\dfrac{2\pi}{T}$, where T is the time for one complete oscillation from peak to peak. ω is therefore equal to 2π times the frequency expressed in cycles per second. But a more general expression for an alternating current must include a phase angle δ, which allows for the possibility of the current not being at zero when $t = 0$. Alternative general expressions in common use are:

$$I = I_0 \sin (\omega t + \delta)$$

and $\quad I = I_0 \cos (\omega t + \delta)$

At this point it must be emphasized that a coil, rotating steadily in a uniform magnetic field, is only an idealized illustration of how an alternating current may be produced. In practice, alternating currents are generated on a large scale in alternators containing rotating and fixed iron-cored coils, but no uniform magnetic field. To a first approximation, the output of such an alternator, and the resulting household electricity supply, may be treated as a sinusoidally-varying e.m.f. Departures from perfect sinusoidal variation may be treated by considering the e.m.f. as a sum of Fourier components which are themselves perfectly sinusoidal: the main component has the principal frequency, while the weaker components are the harmonics with frequencies simple integral multiples of the principal frequency. Thus the British 50 cycle per second grid supply contains harmonics, especially those of frequency 100 and 150 cycles per second.

12·2 The properties of complex numbers

12·1 *Definition of a complex number*

In this section, we describe a mathematical device which will be used throughout the following chapters. The reason for its usefulness, and the method of using it, will be described in section 12·3.

First the symbol j is introduced to denote the operation which, carried out twice in succession, changes the sign of a number, leaving its magnitude unaltered. If x is an ordinary number, jx means x operated on by j; so x operated on twice in succession by j would be written $j(jx)$ or j^2x. Since this has to be equal to $-x$, we see that the double operation denoted by j^2 is equivalent to multiplication by -1. This is what is meant by the statement that j is the 'square root of minus one'. jx is not an ordinary number, and it is called an imaginary number to distinguish it from x, which is 'real'. Real numbers may be added to imaginary numbers to give complex numbers, but the complex number resulting from adding a real number a to an imaginary number jb is nothing more than $a+jb$, two dissimilar quantities added without arithmetical incorporation of the one into the other.

12·2 *The Argand diagram*

There is a convenient graphical device, which allows us to visualize complex numbers, in which real numbers are taken to be coordinates along a line, with positive numbers to the right of a point called the origin, and negative numbers to the left of the origin. Multiplication by -1 is then represented by interchange of the right- and left-hand parts of the line. Such an interchange may be brought about in many ways, one of which is a rotation of the line through 180° about a perpendicular axis through the origin. This rotation may be considered as two successive rotations through 90°. Therefore one such rotation through 90° may be used to represent the operator j. Thus it is possible to represent real numbers on graph paper by distances along the x-axis, while imaginary numbers are represented by distances along the y-axis. A complex number is represented by a two-dimensional vector with x-component equal to its real part and y component equal to its imaginary part. A diagram showing complex numbers in this way is sometimes called an Argand diagram.

Although it has been emphasized that j is an operator, the symbol j is sometimes used alone to mean the operator j acting on 1. In this sense j is represented by a line of unit length along the y-axis of the Argand diagram; and $\frac{1}{j}$ is the same as $-j$, which is represented by a line of unit length along the negative y-axis.

12·2·3 Exponential functions of complex numbers

We now introduce a function of x called exp (jx), which is defined by the relation

$$\exp(jx) = \cos x + j \sin x$$

Another number y will have a corresponding function exp (jy), defined by:

$$\exp(jy) = \cos y + j \sin y$$

If these are multiplied together, and -1 substituted for j^2, we get:

$$[\exp(jx)][\exp(jy)] = (\cos x + j \sin x)(\cos y + j \sin y)$$

$$= (\cos x \cos y - \sin x \sin y)$$

$$+ j(\sin x \cos y + \cos x \sin y)$$

$$= \cos(x+y) + j \sin(x+y)$$

$$= \exp[j(x+y)]$$

(In the fourth step, we used the fact that the right-hand side bore the same relation to $(x+y)$ as our definition makes exp (jx) bear to x.)

Multiplying together, in this way, two identical factors exp (jx), we get:

$$[\exp(jx)][\exp(jx)] = [\exp(jx)]^2 = \exp(j2x),$$

which can be written as exp $(2jx)$. In the same way n equal factors can be multiplied together to give:

$$[\exp(jx)]^n = \exp(njx)$$

This last relation is characteristic of something raised to the power of whatever is inside the bracket, so the function exp (jx) may be given the alternative label e^{jx}. For our purposes it would be sufficient to use the name e^{jx} for the function $\cos x + j \sin x$, remembering that the jx behaves like a power, to be added for multiplication, or to be multiplied by n for raising to the nth power.

However, to make the story more complete, we quote from the mathematical textbooks expressions for $\cos x$ and $\sin x$ as series of terms in ascending powers of x:

$$\cos x = 1 - \frac{x^2}{2!} + \frac{x^4}{4!} - \frac{x^6}{6!} + \ldots$$

$$\sin x = x - \frac{x^3}{3!} + \frac{x^5}{5!} - \ldots$$

These give $\exp(jx) = \cos x + j \sin x$

$$= 1 + jx - \frac{x^2}{2!} - j\frac{x^3}{3!} + \frac{x^4}{4!} + j\frac{x^5}{5!} + \ldots$$

Writing j^2 for -1. and remembering that this requires $j^4 = +1$, we may rewrite this series as:

$$\exp(jx) = 1 + (jx) + \frac{(jx)^2}{2!} + \frac{(jx)^3}{3!} + \ldots + \frac{(jx)^n}{n!} + \ldots$$

which has the same form, with (jx) instead of x, as the standard series for e^x:

$$e^x = 1 + x + \frac{x^2}{2!} + \frac{x^3}{3!} + \ldots + \frac{x^n}{n!} + \ldots$$

Here e is the number obtained by putting $x = 1$:

$$e = e^1 = 1 + 1 + \frac{1}{2!} + \frac{1}{3!} + \ldots$$

$$= 2 \cdot 718 \ldots$$

To summarize by looking back through this argument, we have taken the standard definition of the exponential function e^x, extended it to include exponential functions of imaginary quantities, and shown that e^{jx} so defined is equal to $\cos x + j \sin x$.

2·3 Alternating currents described by means of complex numbers

The use of complex numbers in electricity depends upon the property:

$$e^{j(x+y)} = e^{jx} e^{jy}$$

For example, to an alternating current:

$$I = I_0 \cos(\omega t + \delta),$$

we may add an imaginary term $I_0 j \sin(\omega t + \delta)$, and write:

$$I = I_0 e^{j(\omega t + \delta)}$$

Remembering that this I contains a real part which is the observable current, plus an imaginary part which is included only for purposes of calculation, we may treat it as:

$$I = I_0 e^{j\omega t} e^{j\delta}$$

This has real part:

$$I_0[\cos \omega t \cos \delta + (j \sin \omega t)(j \sin \delta)]$$
$$= I_0 \cos (\omega t + \delta),$$

as required.

This illustrates the particular usefulness of the complex number device, in describing sinusoidally-varying quantities. If sine or cosine notation is being used, a change of phase has to be described by inserting a phase angle δ, which has the effect of altering the amplitude of the component with the original phase, while adding a second component 90° out of phase with it. All this can be very clumsy, in comparison with simple multiplication by a factor $e^{j\delta}$. However, it is always wise, when the complex notation is being used for a new problem, to check our understanding of the problem against that which can be gained from the clumsier, less elegant method.

At this point, a word of warning should be given: complex numbers are not helpful for calculating the power in an A.C. circuit, because the real part of the product VI is not equal to the product of the real parts of V and I (see section 13·7).

Chapter 13
Impedance

3·1   A.C. in a resistance

Let us consider a circuit in which a source of alternating e.m.f. is connected directly to a resistance R. In the complex number notation, the alternating e.m.f. may be represented by:

$$V = V_0 e^{j\omega t}$$

At any moment, the current I is such that RI is equal to the value of E at that moment. This requires

$$I = \frac{V}{R} = \frac{V_0}{R} e^{j\omega t} \tag{13.1}$$

The current is therefore sinusoidally varying, with amplitude $I_0 = \dfrac{V_0}{R}$, and with the same frequency and phase as the e.m.f.

3·2   A.C. in a pure inductance

Now let us connect our source of alternating e.m.f. to a self-inductance L. If the resistance of the circuit is negligible, the e.m.f. of the source must be balanced by the induced e.m.f. $L\dfrac{dI}{dt}$. From this balance the current I can be calculated as follows. If the e.m.f. of the source is:

$$V = V_0 e^{j\omega t},$$

then $L\dfrac{dI}{dt} = V = V_0 e^{j\omega t}$

Splitting the differential coefficient and integrating, we obtain from this:

$$L \int dI = V_0 \int e^{j\omega t} dt$$

which may be integrated to give:

$$I = \frac{V_0}{j\omega L} e^{j\omega t} \tag{13.2}$$

This has real part $\dfrac{V_0}{\omega L}$ sin ωt, which must represent the observable current. We conclude that, when the e.m.f. varies as cos ωt, the current varies as sin ωt, which is 90° behind it in phase. The peak current is $\dfrac{V_0}{\omega L}$.

We could have started with the expression V_0 cos ωt, which being the real part of $V_0 e^{j\omega t}$ represents the observable e.m.f. of the source. Putting this equal to $L\dfrac{dI}{dt}$ and integrating, we should have obtained exactly the same expression for I. This may be taken as evidence that cos $x+j$ sin x does behave like the number e raised to the power jx, in integration as in the operations like multiplication for which we demonstrated their equivalence: or the reader may prefer to write out a direct proof of this equivalence, in integration and in differentiation.

13·3   A.C. in a pure capacity

To demonstrate a third type of load, a resistance-free condenser is connected to our source of alternating e.m.f. In this case, the e.m.f.

$$V = V_0 e^{j\omega t}$$

of the source must be balanced by the potential difference $\dfrac{Q}{C}$ which results from the presence of stored charges $\pm Q$ on the plates of the condenser of capacity C. The charge Q at each moment is therefore given by:

$$Q = CV = CV_0 e^{j\omega t}$$

The current flowing at any moment is the rate of change of Q, which is found by differentiating to be:

$$I = \frac{dQ}{dt} = j\omega C V_0 e^{j\omega t} \tag{13.3}$$

This has real part $-\omega C V_0$ sin ωt, which must represent the observable current; it has a peak value $\omega C V_0$, and phase 90° ahead of that of V.

13·4   Definition of impedance

Equations (13.1), (13.2) and (13.3) all contain, in their right-hand sides, the term $V_0 e^{j\omega t}$. This is just V, the e.m.f. of the source, so the three equations can be equally well written in the form:

$$I = \frac{V}{R}$$

$$I = \frac{V}{j\omega L}$$

$$I = j\omega C V$$

All three equations imply that I is proportional to V, and may be written in the form:

$$I = \frac{V}{Z},$$

or: $\quad ZI = V.$

Z is a multiplying factor, which for a pure resistance is R, for a pure inductance is a complex quantity $j\omega L$, and for a pure capacity is another complex quantity $\frac{1}{j\omega C}$. Z is given the general name of impedance.

We have therefore shown that the alternating current in a component may be related to the e.m.f. applied across it by means of a factor of proportionality, the impedance. For a resistance this impedance is real, so the current and e.m.f. vary with time in the same phase; but for a pure inductance, the imaginary impedance $j\omega L$ is used to describe the fact that the current lags behind the e.m.f. by a quarter-cycle, represented by a phase-angle of 90°. For a pure capacity, the impedance is $\frac{1}{j\omega C}$, which may be written $-\frac{j}{\omega C}$. This also represents a phase-angle of 90° between the current and e.m.f., but in the other sense; so the peak of current in one direction comes quarter of a cycle after the peak of e.m.f. in the *other* direction, which means that it comes quarter of a cycle before the next peak of e.m.f. in the same direction. The conventional description for this state of affairs is that the current leads the e.m.f. by 90°.

By using the expressions for impedance, equation (11.1) can be rewritten in the form

$$\Sigma V = \Sigma Z I \qquad\qquad (13.4)$$

This means that in any closed circuit, the sum of the alternating e.m.f.s of all sources present is equal to the sum of the products impedance × current for all the parts of the circuit. In this form, the equation looks conveniently like the d.c. form of Kirchhoff's second law [equation (3.1)].

The unit of inductance, the henry, is a volt per (ampere/second); since ω is measured in radians per second, or $\sec^{-1}$, an impedance $j\omega L$ will be measured in volts per ampere or ohms. Similarly the unit of capacity, the farad, is a coulomb per volt; therefore an impedance $\frac{1}{j\omega C}$ will be measured in second farad [1] or volt ampere^{-1} again. Impedances of all types whether resistive, inductive or capacitive, are thus measured in ohms.

13·5 A.C. in circuit containing resistance and inductance

If a source of alternating e.m.f. is connected to an inductance and a resistance in series, as in figure 36, then the current I may be calculated by putting:

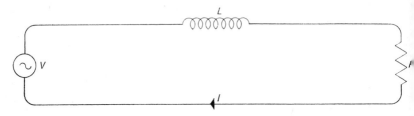

Figure 36. A.C. in circuit containing inductance and resistance.

$$V = RI + j\omega LI$$
$$= (R + j\omega L)I$$

We may obtain this equation directly by saying that the total impedance of R and L in series is:

$$Z = R + j\omega L$$

This is a complex impedance, neither purely real nor purely imaginary, and it can be expressed as a magnitude $|Z|$ multiplied by a phase-factor $e^{j\delta}$, as follows: the ratio imaginary part of Z/real part of Z has to be equal to sin δ/cos δ, which is tan δ. Therefore:

$$\tan \delta = \frac{\omega L}{R},$$

whence:

$$\cos \delta = \frac{1}{\sec \delta} = \frac{1}{\sqrt{(1 + \tan^2 \delta)}} = \frac{1}{\sqrt{\left[1 + \left(\dfrac{\omega L}{R}\right)^2\right]}}$$

$$= \frac{R}{\sqrt{(R^2 + \omega^2 L^2)}}$$

$e^{j\delta}$ must therefore be:

$$e^{j\delta} = \cos \delta + j \sin \delta$$

$$= \frac{R}{\sqrt{(R^2 + \omega^2 L^2)}} + j \frac{\omega L}{\sqrt{(R^2 + \omega^2 L^2)}}$$

$|Z|$ must therefore be equal to $\sqrt{(R^2 + \omega^2 L^2)}$. With:

$$V = V_0 e^{j\omega t}$$

and:

$$Z = |Z| e^{j\delta}$$

the current I must be given by:

$$I = \frac{V}{Z}$$

$$= \frac{V_0 e^{j\omega t}}{|Z| e^{j\delta}}$$

$$= \frac{V_0}{|Z|} e^{j(\omega t - \delta)}$$

This represents an alternating current of peak magnitude $\dfrac{F_0}{\sqrt{(R^2 + \omega^2 L^2)}}$ lagging behind the e.m.f. by a phase-angle $\delta = \tan^{-1} \dfrac{\omega L}{R}$.

·6   A.C. in circuit containing resistance and capacity

In the circuit of figure 37, the total impedance is:

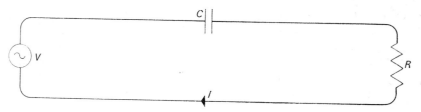

Figure 37.   A.C. in circuit containing capacity and resistance.

$$Z = R + \frac{1}{j\omega C}$$

To express this in the form $|Z| e^{j\delta}$, we must put:

$$\tan \delta = -\frac{1}{\omega C R},$$

and $$|Z| = \sqrt{\left(R^2 + \frac{1}{\omega^2 C^2} \right)}$$

When the e.m.f. is $V_0 e^{j\omega t}$ the current will again be

$$I = \frac{V_0}{|Z|} e^{j(\omega t - \delta)}$$

But this time δ is negative, so the current reaches its peak earlier than does the e.m.f. Thus the current is said to lead the e.m.f. by a phase-angle

$$-\delta = \tan^{-1}\left(\frac{1}{\omega CR}\right)$$

this has already been shown to be $90°$ for a circuit with $R = 0$ and $C \neq 0$.

13·7 Phase-angle and power

If a source of e.m.f. $V = V_0 e^{j\omega t}$ is causing a current I to flow in a load with complex impedance $Z = |Z| e^{j\delta}$, the rate at which the source is providing energy at any moment is given by VI. Here it is necessary to start with the actual observable values, rather than the complex quantities (see warning at the end of section 12·3). The instantaneous rate of providing energy is:

$$VI = V_0 \cos \omega t \frac{V_0}{|Z|} \cos (\omega t - \delta)$$

$$= \frac{V_0^2}{|Z|} \cos \omega t \cos (\omega t - \delta)$$

$$= \frac{V_0^2}{2|Z|} [\cos (2\omega t - \delta) + \cos \delta]$$

The first term in this square bracket is oscillatory and represents an oscillatory flow of energy alternately out of the source and back into it. However, super imposed on this oscillatory flow of energy there is a steady flow from the source represented by the second term in the square bracket, $\cos \delta$. The average rate of supply of energy by the source is therefore equal to:

$$\frac{V_0^2}{2|Z|} \cos \delta$$

If I_0, which is the peak current, is substituted for $\frac{V_0}{|Z|}$, this becomes:

$$\tfrac{1}{2} V_0 I_0 \cos \delta$$

Further simplification is obtained by using root-mean-square e.m.f. and current: if bars are used to indicate mean values, $V_{\text{r.m.s.}}$ is defined by:

$$V_{\text{r.m.s.}} = \left(\overline{V^2}\right)^{\frac{1}{2}}$$

$$= \left(\overline{V_0^2 \cos^2 \omega t}\right)^{\frac{1}{2}}$$

$$= V_0 \left\{ \frac{1}{2}(\overline{1 + \cos 2\omega t}) \right\}^{\frac{1}{2}}$$

$$= \frac{V_0}{\sqrt{2}}$$

since the mean value of $\cos 2\omega t$ is zero.

Similarly:

$$I_{\text{r.m.s.}} = \frac{I_0}{\sqrt{2}}$$

The two factors $\dfrac{1}{\sqrt{2}}$ can make up the $\dfrac{1}{2}$ in $\dfrac{1}{2}V_0 I_0 \cos \delta$. Therefore the average rate of supply of energy by the source, which is called the power supplied by the source, may be written as:

$$V_{\text{r.m.s.}} \, I_{\text{r.m.s.}} \, \cos \delta$$

Therefore if we describe alternating e.m.f.s and currents by their root-mean-square values, as is customary, the power in a circuit is obtained by multiplying them together, with an extra factor $\cos \delta$, known as the power factor.

Chapter 14
Resonance

14·1 Series resonant circuit

14·1·1 Derivation of Q-factor

By an extension of the discussion of section 13·5, let us consider the case of a circuit containing a source of alternating e.m.f., with resistance, inductance and capacity in series, as in figure 38.

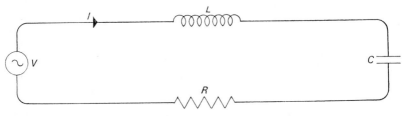

Figure 38.   Series resonant circuit.

The total impedance in this circuit is:

$$Z = R + \frac{1}{j\omega C} + j\omega L = R + j\left(\omega L - \frac{1}{\omega C}\right)$$

which may be expressed as $|Z|\ e^{j\delta}$ if:

$$|Z| = \sqrt{\left[R^2 + \left(\omega L - \frac{1}{\omega C}\right)^2\right]}$$

and $\tan \delta = \dfrac{1}{R}\left(\omega L - \dfrac{1}{\omega C}\right)$

The interesting point about this circuit is that there exist values of ω, L and C which make Z purely real, with $\tan \delta = 0$ and $|Z| = R$. These are the values which make:

$$\omega L = \frac{1}{\omega C}$$

$$\omega = \frac{1}{\sqrt{(LC)}}$$

$\dfrac{1}{\sqrt{(LC)}}$ is a characteristic angular frequency of the circuit; it is called the resonant angular frequency, and is denoted by the symbol ω_0.

For a given L and C, $|Z|$ and δ vary with ω as shown in figures 39 and 40. At the

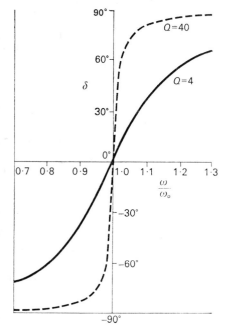

Figure 39. Phase-angle in series resonant circuit.
Full line: $Q = 4$
Broken line: $Q = 40$

resonant frequency $|Z|$ reaches a minimum value of R, and δ goes through zero. Below the resonant frequency δ is negative, the circuit behaving like a capacitative load; but above the resonant frequency δ has a positive value, which is characteristic of an inductive load.

At the resonant frequency, $\omega = \omega_0$ and L, C and R together behave like R alone, giving current:

$$I = \frac{V}{R}$$

But this current must flow through L, and there will be an induced e.m.f. across the inductance equal to:

$$V_L = j\omega LI$$

Similarly there will be a potential difference across the condenser, equal to:

$$V_C = \frac{I}{j\omega C} = -\frac{jI}{\omega C}$$

Since ωL is equal to $\dfrac{1}{\omega C}$ when $\omega = \omega_0 = \dfrac{1}{\sqrt{(LC)}}$, V_C and V_L exactly cancel each other. Both are 90° out of phase with the current and the magnitude of each is given by:

$$\frac{1}{j} V_L = V_C = \omega_0 LI = \frac{\omega_0 L}{R} V$$

The dimensionless factor $\dfrac{\omega_0 L}{R}$ may alternatively be written $\dfrac{1}{\omega_0 CR}$ or $\dfrac{1}{R}\sqrt{\left(\dfrac{L}{C}\right)}$ it is usually given the symbol Q and is called the 'Q' or Q-factor of the circuit. It should be noted that if R is small, Q may be a large number; in this case, the e.m.f.s across L and C are correspondingly larger than the e.m.f. of the source

14·1·2 Use of Q-factor

Q is useful also for describing the frequency-dependence of Z near the resonance This can be described most easily if the imaginary parts of Z are written in terms of Q and ω_0 as follows:

$$Z = R + j\left(\omega L - \frac{1}{\omega C}\right)$$

$$= R\left\{1 + jQ\sqrt{\frac{C}{L}}\left(\omega L - \frac{1}{\omega C}\right)\right\}$$

$$= R\left\{1 + jQ\left(\frac{\omega}{\omega_0} - \frac{\omega_0}{\omega}\right)\right\}$$

The factor $\left(\dfrac{\omega}{\omega_0} - \dfrac{\omega_0}{\omega}\right)$ is large and negative when ω is small; it reaches zero when ω becomes equal to ω_0, and is large and positive when ω is large. However for frequencies which differ from ω_0 by a small fraction a, we may write:

$$\omega = \omega_0(1+a),$$

and: $\dfrac{\omega}{\omega_0} - \dfrac{\omega_0}{\omega} = 1 + a - \dfrac{1}{1+a} \approx 2a$

At such frequencies therefore:

$$Z = R(1 + 2jQa)$$

This reminds us that Z has a real part equal to R at all frequencies; it shows also that Z has an imaginary part proportional to the fraction a by which ω differs from ω_0. The imaginary part of Z becomes equal in magnitude to the real part at frequencies which have:

$$a = \pm \frac{1}{2Q}$$

At these frequencies the phase-angle δ has values $\pm 45°$, and $|Z|$ has the value $\sqrt{2}R$. Q thus measures the steepness of the plot of δ against frequency near the resonance, and the sharpness of the trough containing the minimum value of $|Z|$. Typical curves, for large and small values of Q, are shown in figures 39 and 40.

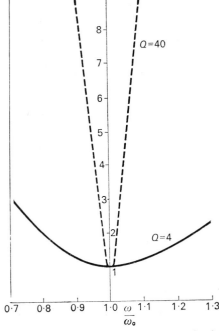

Figure 40. $|Z|/R$ for series resonant circuit.
Full line: $Q = 4$
Broken line: $Q = 40$

The above discussion shows that a series resonant circuit may be used as a filter, for when a source of e.m.f. generating several frequencies is connected to it, the lowest impedance will be offered to the component with the resonant frequency. Therefore the current which flows through the filter will contain a greater proportion of the component with this frequency than did the e.m.f. Complete filtering is not obtained, but components with frequency a long way from resonance may be severely attenuated.

14·2   Parallel resonant circuit

An alternative way of connecting an inductance and a capacity is shown in figure 41. In this case they are connected in parallel, so that different currents can flow

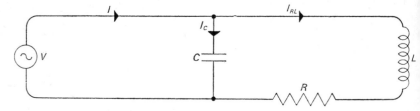

Figure 41.   Parallel resonant circuit.

through each. To be realistic, we have to include some resistance in the arm containing L; but for convenience the small resistance associated with C can be ignored.

The total current I will be just the sum of the currents I_C and I_{RL} through the two arms. Using the symbols Z_C and Z_{RL} to represent the impedances of the two arms, we may write:

$$I = I_C + I_{RL}$$

$$= \frac{V}{Z_C} + \frac{V}{Z_{RL}}$$

$$= V.j\omega C + \frac{V}{R+j\omega L}$$

$$= V\left\{ j\omega C + \frac{1}{(R+j\omega L)} \right\}$$

At this point ratio $\dfrac{V}{I}$ may be equated with Z, the effective impedance of the circuit; but in fact for a parallel circuit, it is more convenient to use its reciprocal, which is called the admittance Y of the circuit. The admittance is the a.c. analogue of conductance and in this case is:

$$Y = \frac{1}{Z} = \frac{I}{V} = j\omega C + \frac{1}{(R+j\omega L)}$$

$$= j\omega C + \frac{R-j\omega L}{(R^2 + \omega^2 L^2)}$$

$$= \frac{R}{(R^2 + \omega^2 L^2)} + j\omega \left\{ C - \frac{L}{(R^2 + \omega^2 L^2)} \right\}$$

The algebra of reducing this expression is rather laborious. But in the case $\omega L \geqslant R$ the conclusions, which may be checked by the student, are as follows:

(i) The imaginary term in Y becomes zero at a frequency close to the frequency $\omega_0 = \dfrac{1}{\sqrt{(LC)}}$ which would be the resonant frequency for the same components arranged as a series resonant circuit.

(ii) The magnitude $|Y|$ of the admittance reaches a minimum value at a frequency very close to ω_0. The circuit therefore acts as a filter in the sense opposite to the series resonant circuit: since $|Y|$ is a minimum at resonance, $|Z|$ is a maximum, and a given applied e.m.f. causes minimum current to flow. Correspondingly a given current causes a maximum alternating e.m.f. to be set up across the circuit, when the frequency is ω_0.

The actual value of $|Z|$ at resonance is $RQ\sqrt{(1+Q^2)}$, where Q is the parameter defined for the series resonant circuit by:

$$Q = \frac{1}{R}\sqrt{\left(\frac{L}{C}\right)} = \frac{\omega_0 L}{R} = \frac{1}{\omega_0 CR}$$

For small R, Q is large, and the maximum $|Z|$ is approximately $Q^2 R$.

(iii) At frequencies near ω_0 we may adopt the device used for the series resonant circuit, and obtain closely-related results:

If $\qquad \omega = \omega_0(1+a)$

Y is given, to a close approximation, by:

$$Y = \frac{1}{RQ\sqrt{(1+Q^2)}}(1+j2Qa)$$

Thus $|Y|$ exceeds its minimum value by a factor $\sqrt{2}$ when:

$$a = \pm\frac{1}{2Q}$$

At the frequencies which satisfy this condition, one on each side of ω_0, the phase-angle is $45°$, and Y and Z have imaginary parts equal to their real parts.

4·3 Conditions at resonance

This section is a recapitulation, with changed emphasis, of some salient features of the preceding discussion.

When an alternating e.m.f. of the resonant frequency is applied to a series resonant circuit, there are set up across the condenser and the inductance e.m.f.s which are $90°$ out of phase with the applied e.m.f. and $180°$ out of phase with each other; these both have magnitude Q times that of the applied e.m.f. If it should happen that Q is large, say 30 or 40, the alternating e.m.f.s across the condenser and the inductance may be surprisingly high. For example, if resonance is liable to occur in a low-resistance LC circuit connected to a 200-volt alternating source, the condenser should be rated to stand a peak voltage of several kilovolts (6 KV. if $Q = 30$).

A similar effect occurs in the parallel resonant circuit, but here it is a large alternating current, which flows round the LC loop, 90° out of phase with the e.m.f. of the source and the small current which flows from it. Physically we can see that this current will have the same magnitude as would the current through one of the components if it alone were connected to the source, namely $\omega_0 C V_0$,

or $\dfrac{V_0}{QR}$. This is larger by a factor of approximately Q than the peak current

$\dfrac{V_0}{QR\sqrt{(1+Q^2)}}$ which flows through the parallel resonant circuit considered as a whole. Therefore, just as the condenser in a series resonant circuit has to be rated to stand a peak voltage Q times larger than the peak voltage of the source, the inductance in a parallel resonant circuit must be made of wire thick enough to stand a current Q times that which is to be drawn from the source.

14·4 Sharpness of resonance

Either type of LC circuit may be used as a filter. One of the commonest such filters is the tuning circuit of an ordinary radio receiver; this is a parallel resonant circuit, connected between earth and a wire carrying the signals from the aerial. Many signals are present, each taking the form of an alternating current of characteristic frequency. To most of these the parallel LC circuit offers small impedance, so that the e.m.f. set up across it is small. However, if there is a signal of frequency close to ω_0, resonance will occur, and a larger e.m.f. will be set up. This signal will thus have been selected from the others; the choice may be varied by varying the capacity of the condenser. The selection is, however, not infinitely sharp, because a signal of any frequency within the limits $\omega_0\left(1\pm\dfrac{1}{2Q}\right)$ will give

an e.m.f. within a factor $\dfrac{1}{\sqrt{2}}$ of the peak value which it would give if it had frequency exactly ω_0. A filter of this simple form is thus effective at separating signals whose frequencies differ by amounts of not less than about $\dfrac{\omega_0}{Q}$.

Chapter 15
The transformer

Simple theory

A transformer is essentially a pair of coils wound in such a way that they have the largest possible mutual inductance; this usually means that they are wound on a common ferromagnetic core. In an ideal case, the whole of the magnetic flux resulting from a current in one of the coils will pass through both coils.

Let us imagine an ideal transformer, as sketched in figure 42, with two coils

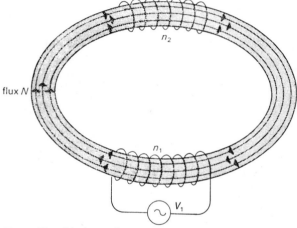

Figure 42.   Ideal transformer.

wound on a ring-shaped core in which there is a flux N, the coils having numbers of turns n_1 and n_2 respectively. If the flux is changing, there will be induced in the coils e.m.f.s equal to $\dfrac{dN}{dt}$ per turn. The total induced e.m.f. in coil 1 is therefore of magnitude:

$$n_1 \frac{dN}{dt},$$

while that in coil 2 is:

$$n_2 \frac{dN}{dt}$$

Let us now suppose that the flux N results from the flow of an alternating current in coil 1, which is connected to a source of alternating e.m.f. V_1. If there is no resistance in the circuit, the e.m.f. of the source must exactly balance the induced e.m.f. in the coil. The condition for this is:

$$V_1 = n_1 \frac{dN}{dt}$$

It follows that the induced e.m.f. in coil 2 is:

$$V_2 = n_2 \frac{dN}{dt} = n_2 \frac{V_1}{n_1} = \frac{n_2}{n_1} V_1 \qquad (15.1)$$

The apparatus has thus 'transformed' the alternating e.m.f. V_1 to another which differs from it by a factor $\frac{n_2}{n_1}$. This is the ratio of the numbers of turns on the two coils.

This very simple theory holds for the case of perfect flux linkage (i.e. the same N and $\frac{dN}{dt}$ for both coils), so long as no current flows in coil 2, which we may now call the secondary coil. Under this condition, the primary coil behaves like a perfect self-inductance, with current I given by:

$$V = j\omega L I$$

This is consistent with our earlier statement that the flux N was due to (and presumably proportional to) the current I, while V was proportional to $\frac{dN}{dt}$.

15·2 Fuller theory of ideal transformer

15·2·1 *Extension of simple theory*

Let us now examine the balance of e.m.f.s in a transformer, when current is allowed to flow in the secondary circuit. In figure 43, the transformer is drawn as a mutual inductance M, the two coils having also self-inductances L_1 and L_2. The primary coil is connected to a source of alternating e.m.f. V, while the secondary coil is connected to a load of impedance Z. For convenience, the

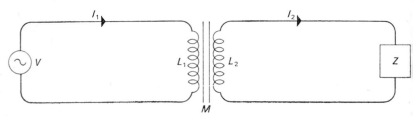

Figure 43. Transformer circuit.

resistances of the coils are assumed to be included in the source and in Z respectively.

A current I_1 in the primary circuit sets up induced e.m.f.s:

$$j\omega L_1 I_1 = L_1 \frac{dI_1}{dt} \text{ in the primary circuit, and}$$

$$j\omega M I_1 = M \frac{dI_1}{dt} \text{ in the secondary circuit.}$$

Similarly, a current I_2 in the secondary circuit sets up induced e.m.f.'s:

$$j\omega L_2 I_2 = L_2 \frac{dI_2}{dt} \text{ in the secondary circuit, and}$$

$$j\omega M I_2 = M \frac{dI_2}{dt} \text{ in the primary circuit.}$$

Collecting these up, we obtain for the balance of e.m.f.'s in the two circuits:

$$V = j\omega L_1 I_1 + j\omega M I_2 \tag{15.2}$$

$$0 = j\omega M I_1 + j\omega L_2 I_2 + Z I_2 \tag{15.3}$$

These may be solved as two simultaneous equations for two unknowns, I_1 and I_2, in terms of V and the constant factors. The imaginary nature of some of the factors does not alter the method of solution, which may follow the ordinary methods of elementary algebra, as follows.

Equation (15.3) may be reorganized to give I_2 in terms of I_1:

$$I_2 = -\frac{j\omega M}{j\omega L_2 + Z} I_1 \tag{15.4}$$

Substituting equation (15.4) in (15.2), gives I_1 in terms of V:

$$V = I_1 \left\{ j\omega L_1 + j\omega M \left(\frac{-j\omega M}{j\omega L_2 + Z} \right) \right\}$$

$$= I_1 \left\{ \frac{-\omega^2 L_1 L_2 + j\omega L_1 Z + \omega^2 M^2}{j\omega L_2 + Z} \right\}$$

$$= I_1 \left\{ \frac{\omega^2 (M^2 - L_1 L_2) + j\omega L_1 Z}{j\omega L_2 + Z} \right\}$$

$$\therefore I_1 = V \left\{ \frac{j\omega L_2 + Z}{\omega^2 (M^2 - L_1 L_2) + j\omega L_1 Z} \right\} \tag{15.5}$$

Substituting this in equation (15.4) gives I_2 as a function of V:

$$I_2 = V \left\{ \frac{-j\omega M}{\omega^2 (M^2 - L_1 L_2) + j\omega L_1 Z} \right\} \tag{15.6}$$

15·2·2 Limiting value of M

Equations (15.5) and (15.6) contain most of the information needed for discussing the operation of a transformer; they contain also evidence about the theoretical upper limit of M for given values of L_1 and L_2. If the impedance Z of the load is allowed to become small the denominator in each of the equations becomes $\omega^2(M^2-L_1L_2)$; if this were zero, the currents would be infinite. When the coils are not very closely linked, M is small, making $(M^2-L_1L_2)$ negative; then as the coils become more and more closely linked, M becomes larger, and $(M^2-L_1L_2)$ becomes smaller in magnitude; but however closely we link the coils, we can never quite reach the point at which:

$$M^2 = L_1L_2$$

Nor can we go beyond this point, because positive $(M^2-L_1L_2)$ would imply that the whole system had impedance like a negative inductance and could act as a source of energy.

This means that the upper limit to M is:

$$M = \sqrt{(L_1L_2)}$$

as mentioned in Chapter 7.

15·2·3 Application to ideal transformer

Having shown that M can never be quite equal to $\sqrt{(L_1L_2)}$, we now turn a mental somersault and take as our ideal transformer one in which M is effectively equal to its limiting value. In other words having used the case of indefinitely small Z for its theoretical implication, we discard it as having less physical interest than the case of infinitely small $(M^2-L_1L_2)$. In this ideal case, equations (15.5) and (15.6) become:

$$I_1 = V\frac{j\omega L_2+Z}{j\omega L_1 Z} = V\left(\frac{L_2}{L_1 Z}+\frac{1}{j\omega L_1}\right) \tag{15.7}$$

$$I_2 = V\frac{-j\omega M}{j\omega L_1 Z} = -V\frac{M}{L_1 Z} \tag{15.8}$$

Equation (15.8) contains several facts of direct physical significance:

(i) For a resistive load with real Z, I_2 is in phase with V; for any load, resistive or otherwise, I_2 has the same phase as it would if the load were connected directly to the source.

(ii) The magnitude of I_2 differs from the value $\dfrac{V}{Z}$, which would be obtained by connecting the load directly to the source, by a factor $\dfrac{M}{L}$. If the value $\sqrt{(L_1L_2)}$ is used for M, this factor becomes $\sqrt{\left(\dfrac{L_2}{L_1}\right)}$.

(iii) The e.m.f. across the secondary coil and across the load, is just I_2Z, which is equal to $\sqrt{\left(\dfrac{L_2}{L_1}\right)}\,V$. It was mentioned in section 7·2 that the self-inductance of a coil was proportional to the square of the number of turns on it; $\sqrt{\left(\dfrac{L_2}{L_1}\right)}$ is therefore just the ratio $\dfrac{n_2}{n_1}$ of the numbers of turns on the two coils. It has thus been shown that the equation (15.1) for the e.m.f. across the open-circuited secondary coil of a transformer holds for an ideal transformer whether or not it is on open circuit.

The facts contained in equation (15.7) are discussed in section 15·3.

·3 Effect of secondary load on primary current

When the load impedance Z is large, equation (15.7) shows that I_1 becomes practically equal to $\dfrac{V}{j\omega L_1}$; this is the behaviour characteristic of a simple self-inductance, which it was indicated in section 15·1 would be expected when the secondary coil was on open circuit (i.e. Z infinite).

But we now see that the effect of decreasing Z is not only to increase I_2, but also to add to I_1 an increasingly large component with the phase of $\dfrac{V}{Z}$. When Z is the real impedance of a resistive load this component is in phase with V, unlike the component $\dfrac{V}{j\omega L_1}$ which is 90° out of phase with E.

This in-phase component of primary current gives us a clue to the mechanism whereby the power dissipated in the load is extracted from the source. When no current is flowing in the secondary circuit, the primary current is not zero, but is 90° out of phase with E and therefore draws no power from the source. However, when a resistive load R is connected, power is dissipated in it at a rate equal to $R \times$ (mean value of I_2^2). The student is advised to check that this is in fact equal to the power $E_{\text{r.m.s.}}.I_{\text{r.m.s.}}.\cos\,\delta$ which is drawn from the source when the phase-angle δ is not 90° but:

$$\tan^{-1}\left(\frac{-1}{\omega L_1}\bigg/\frac{L_2}{L_1 Z}\right)$$

$$= \tan^{-1}\left(\frac{-Z}{\omega L_2}\right)$$

Insofar as transformers are ideal this transfer of power is efficient; we may therefore leave the primary coil of a transformer connected to an alternating supply, knowing that power will be drawn from the source only when a power-consuming load is connected to the secondary coil. The most familiar example of this is the small transformer for household electric bells. Some of these are far from ideal, but leaving them connected still adds a negligible amount to the bill for power consumed.

15·4 Real transformers

There are three principal points in which real transformers depart from the ideal discussed above; none of them is important so far as the secondary current is concerned, but we shall discuss the effect which they have on the current and power in the primary circuit.

(i) *Imperfect flux linkage*
If some of the flux produced by the primary coil fails to go through the secondary, the mutual inductance M will be less than its limiting value $\sqrt{(L_1 L_2)}$. In such circumstances, the term $\omega^2(M^2 - L_1 L_2)$ in the denominator of equations **(15.5)** and **(15.6)** becomes a real negative quantity, which is not negligibly small. The algebra of handling the equations with this term included is clumsy; but it may be shown algebraically, as it can be seen physically, that a transformer with imperfect flux linkage is equivalent to a perfect transformer with a small series self-inductance in the primary circuit.

(ii) *Eddy currents*
Transformer cores are usually laminated, to restrict the flow of induced current in the core itself. However, some such induced currents will always occur. These are called eddy currents. Their effect on the primary and secondary circuits may be described by considering them to be currents flowing in additional short-circuited secondary turns. The primary current therefore has an extra component in phase with V; extra power is thus drawn from the source, as if there were an extra resistive load connected in parallel with the actual load Z.

(iii) *Hysteresis*
The material of a transformer core is being continually taken round a magnetic cycle. Its magnetization traces out a hysteresis loop, and as it was pointed out in Chapter 8 the area of such a loop represents the energy dissipated per cycle in the material. In a transformer, the material of the core is chosen to have a small hysteresis loss, but whatever energy is lost in this way must be provided by the source. The mechanism for this is yet another small component of current, in phase with V.

Chapter 16
A.C. measurements

Measurement of current

·1   *The moving-coil meter*

A direct current is normally measured with the aid of a moving-coil meter. This consists of a rectangular coil mounted on pivots in the field of a permanent magnet, so that it can move in response to the couple which is illustrated in figure 13 and discussed in Chapter 6. The movement is resisted by a hairspring; the restoring couple due to the hairspring is proportional to the angular displacement, and the twisting couple due to the current is proportional to the current. In equilibrium the coil takes up a position in which these couples balance. The angular displacement of the equilibrium position may therefore be taken as being proportional to the current, and vice versa when the current is being deduced from the position of a pointer attached to the coil. In fact the couple is proportional to the current only if the component of B in the plane of the coil is independent of the position of the coil. This condition cannot be achieved exactly, but a close approximation is obtained for displacements over a range 45° to 60°, by cylindrical shaping of the pole-pieces. If an iron cylinder is fixed between the pole-pieces, coaxially with their shaping, the coil can swing in a nearly-uniform radial field in the narrow space between the cylinder and the pole-pieces.

Such meters are usually fitted with external resistances, which are chosen so that a particular mechanism may be used for different purposes. In ammeters and milliammeters parallel resistances serve to carry part of the current to be measured, leaving a known fraction to go through the coil. In voltmeters, large series resistances are used to limit the current due to a given voltage to a value within the range of the mechanism. The exact arrangement of series and parallel resistances chosen for a particular use must not only give the desired sensitivity, but also give reasonable damping: when the coil is moving, an induced e.m.f. is set up, and current tends to flow in the direction which slows down the movement. The magnitude of this induced current, and of the damping which results from it, is inversely proportional to the resistance of the external circuit through which it must flow. On open circuit, most meter mechanisms are under-damped, the coil swinging to and fro many times before reaching equilibrium; but when short-circuited they are over-damped, creeping slowly towards a final position. Critical damping, with reasonably fast approach to the final position but no overshooting, is achieved by use of a suitable external resistance. In many meters, part of the necessary damping is provided by winding the coil on a metal former, which acts as a short-circuited single-turn coil.

16·1·2   *Meters for measuring a.c.*

If an alternating current is passed through an ordinary moving-coil meter no deflection is observed, because during the half-cycle in which the current is in one direction, the torque on the coil will be in one sense, but during the next half-cycle, with the current flowing the other way, the torque will be in the other sense. If the coil and pointer are extremely light they may be observed to vibrate a little with the frequency of the alternating current, but there is no net displacement.

It is possible to construct meters which give deflections in one direction only, regardless of the direction of the current. Examples of this are hot-wire instruments, and moving-iron instruments, in which a stationary coil acts like a solenoid, magnetizing and drawing into itself a piece of iron. In the modern version of the hot-wire instrument, the small temperature rise of a wire carrying the current in question is measured by means of a thermocouple; the latter is connected to an ordinary moving-coil millivoltmeter. All these unidirectional instruments give deflections (approximately) proportional to the square of the current, when it is steady, or to the mean square when it is alternating. Therefore they have non-linear scales, tolerable in some circumstances, but inconvenient for many uses. If they are calibrated with direct current, the calibration gives the root-mean-square value of an alternating current which causes the same deflection.

16·1·3   *Rectification of a.c.*

A very convenient way of measuring an alternating current, much used in portable instruments, is to rectify it and measure the resulting direct current. Silicon rectifiers are obtainable in conveniently small sizes, with low resistance for currents in one direction, and resistance many orders of magnitude larger for currents in the reverse direction. Such rectifiers are often made in sets of four, which can be arranged in a bridge circuit (see figure 44); this allows full-wave rectification, with each half-cycle of the alternating current flowing through the

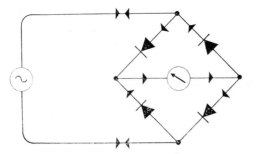

Figure 44.   Rectifier bridge for full-wave rectification.

meter in the chosen direction. The current through the meter is not steady, but varies with time as shown in figure 45. If the meter is an ordinary moving-coil

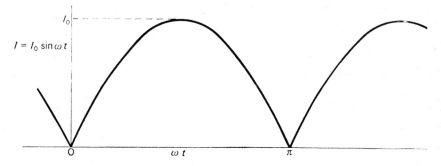

Figure 45.   Rectified alternating current.

type, its coil and pointer will take up a position such that the restoring torque of the hairspring balances the mean torque due to the current. This mean torque is proportional to the mean rectified current, which is the arithmetic mean alternating current, *not* to the root-mean-square current. The relation between these is found by integrating over a quarter-cycle. If the current is:

$$I = I_0 \sin \omega t,$$

the arithmetic mean current is:

$$\bar{I} = \frac{\displaystyle\int_0^{\frac{\pi}{2}} I d(\omega t)}{\displaystyle\int_0^{\frac{\pi}{2}} d(\omega t)}$$

$$= I_0 \frac{\left[-\cos \omega t \right]_0^{\frac{\pi}{2}}}{\left[\omega t \right]_0^{\frac{\pi}{2}}}$$

where square brackets indicate that the quantity inside must be taken over the limits specified as upper and lower suffixes. This gives the ratio:

$$\frac{\text{arithmetic mean current}}{\text{peak current}} = \frac{\bar{I}}{I_0} = \frac{1}{\dfrac{\pi}{2}} = \frac{2}{\pi} = 0.6366 \ldots$$

The ratio of the arithmetic mean current to the root-mean-square current is thus:

$$\frac{\bar{I}}{I_{\text{r.m.s.}}} = \frac{2\sqrt{2}}{\pi} = 0.9003 \ldots$$

Most a.c. instruments are intended for use with sinusoidally-varying currents, and are required to measure r.m.s. currents. The scales are therefore calibrated to indicate, not the direct current which would give the same deflection, but a current $\dfrac{\pi}{2\sqrt{2}}$ ($= 1\cdot11$ approx.) times larger. This gives the r.m.s. value of the alternating current causing the given deflection; but in the rare cases when such an instrument is connected to a source of oscillations which are not sinusoidal (e.g. a square-wave or a saw-tooth generator), it will give scale readings equal to $1\cdot1107$ times the arithmetic mean current. This may be very different from the r.m.s. current.

16·1·4 *Measurement of rectified a.c.*

In order to measure current or voltage in a.c. circuits, a moving-coil meter with its rectifier unit may be fitted with a variety of parallel or series resistors, to give different ranges of sensitivity, just as can the simple meter in a d.c. instrument. If the same meter is to be used for both a.c. and d.c. measurements, a single scale calibration can cover both, provided two conditions are satisfied. The first is that the external resistances for a.c. use must be calculated to give the required sensitivity when the effective resistance of the rectifier unit is added to that of the meter mechanism; they will therefore not be the same as those required for d.c. use. The second is that the rectifiers must offer the same resistance to large currents as they do to small ones in the same direction, over the required range. In ammeters for a.c. only, a transformer may be used instead of a resistance network for range-selection.

16·2 Measurement of impedance

16·2·1 *The Wheatstone bridge*

Resistances may be conveniently, and accurately, measured by means of d.c. bridge circuits. The simplest of these is the Wheatstone bridge, which appears in different guises in various instruments. In principle this passes current from a given source through a potential divider consisting of two known resistances while a second current from the same source flows through another potential divider consisting of one known and one unknown resistance. One (or two) of the known resistances is adjusted until the centre points of the two potential dividers are at the same potential; this is indicated by zero deflection on a sensitive galvanometer connected between them.

The same principle may be applied to an alternating current bridge circuit, for comparing impedances with inductive or capacitive as well as resistive components. A general circuit of this sort is illustrated in figure 46.

The impedances Z_1, Z_2, Z_3 and Z_4 are unspecified for the moment; the source of alternating e.m.f. may be an oscillator of any convenient frequency, say 1000 cycles per second. The detector D does not need to measure currents quantita-

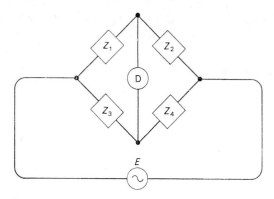

Figure 46. General A.C. bridge.

tively, but only has to indicate the approach toward zero current. It is sometimes convenient to use headphones as detectors when the frequency is in the audible range, but the accuracy of this method is limited by the stray capacities which it introduces, and harmonics which can be confusing. More systematic measurements are possible if we use as detector a sensitive moving-coil meter, incorporating a rectifier as discussed in section 16·1. It must be remembered however, that this will always give a deflection in the same direction; the pointer does not move from one side of zero to the other on passing through a balance-point. In the best instruments, an amplifier is included to give a more sensitive indication of balance.

The circuit of figure 46 is just a Wheatstone bridge circuit. It will give a balance when the ratio Z_1/Z_2 of the upper impedances is equal to the ratio Z_3/Z_4 of the lower impedances in both magnitude and phase; inequality of either the magnitudes or the phases will lead to the observation of an alternating current in the detector. There are thus two conditions which must be satisfied before an a.c. bridge is balanced. We may extract many alternative pairs of conditions by treating the equations in different ways. All such pairs of conditions are equivalent, so long as they satisfy the complex equation:

$$\frac{Z_1}{Z_2} = \frac{Z_3}{Z_4}$$

Instead of considering the magnitudes and phases in this equation, we shall write the equivalent equation

$$Z_1 Z_4 = Z_2 Z_3 \tag{16.1}$$

and say that the real parts of the two sides must be equal, as must the imaginary parts. Let us write

$$Z_1 = Re_1 + jIm_1$$

where Re_1 means the real part of Z_1 and Im_1 means the imaginary part of Z_1; with Z_2, Z_3 and Z_4 similarly represented, the conditions for balance become:

$$Re_1 Re_4 - Im_1 Im_4 = Re_2 Re_3 - Im_2 Im_3 \qquad (16.2)$$

and: $$Re_1 Im_4 + Re_4 Im_1 = Re_2 Im_3 + Re_3 Im_2 \qquad (16.3)$$

The first of these comes from the real part of equation (16.1), while the second comes from its imaginary part. Together they constitute the conditions for balance in a general bridge circuit.

Particular cases of the simple bridge circuit, with two arms purely resistive, may be used for special purposes, e.g. comparing capacities in adjacent arms, inductances in adjacent arms, or an inductance and a capacity in opposite arms. But instead of going into the details of these and other special circuits we shall discuss another much used type of bridge circuit.

16·2·2 Transformer ratio arm bridge

Consider the circuit shown in figure 47, in which one transformer provides two

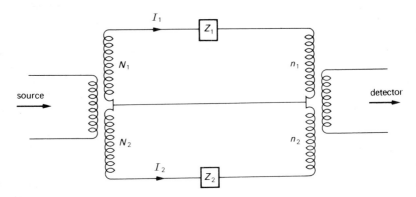

Figure 47.   Transformer ratio-arm bridge.

voltages in the ratio of the numbers of turns of its two secondary windings, say $r = N_1/N_2$. These two voltages produce currents I_1 and I_2 in the impedances Z_1 and Z_2, and the currents go through separate primary windings in a second transformer. If the windings have numbers of turns in the ratio $r' = n_1/n_2$, the currents cause e.m.f.s in the secondary with ratio $r' I_1/I_2$. If the windings are arranged in senses such that these induced e.m.f.s cancel, zero output voltage can be observed in the detector when the impedances satisfy the condition

$$Z_2 = rr' Z_1 \qquad (16.4)$$

Z_1 and Z_2 should include the impedances of the transformer windings but in normal conditions these can be negligibly small, say of order 100 Ω for comparing two impedances of order 10 to 100 MΩ.

With transformers of suitable characteristics, and with the possibility of choosing different pairs of windings, bridges of this type have been used as the basis of many flexible instruments for comparing capacities and inductances. Usually the resistive part of the unknown impedance is regarded as an imperfection, to be balanced by a continuously variable resistance in the opposite arm while the inductive or capacitive parts are compared quantitatively.

3 Measurements of inductance and capacity

Pairs of capacities or inductances may be compared with a simple bridge or a transformer ratio-arm bridge. In the case of capacities special problems arise from the difficulty of defining the capacity required. These problems, and the techniques for making accurate measurements despite them, merit some consideration.

In principle the capacity between two conductors is definite only if one of the conductors completely surrounds the other; otherwise it depends on the configuration of other conductors in the neighbourhood. In practice two conductors can have a definite capacity if a third conductor is introduced to enclose one of them completely, or enclose both nearly enough to ensure that their terminals are shielded from each other.

For accurate work, one thus has the "three-terminal capacitor", with two terminals for the active conductors between which the direct capacitance is defined, and a third terminal for a screen with which the active conductors have ill-defined but maybe large capacitances. Circuits must be devised to compare direct capacitances in the presence of these ill-defined additional capacitances.

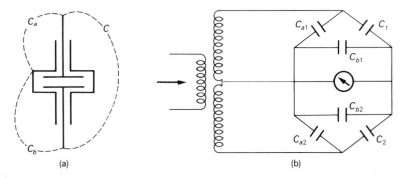

Figure 48. (a) Three-terminal capacitor.
(b) Bridge for comparing capacitances.

In the example shown in figure 48, balance is achieved when the direct capacities C_1 and C_2 are the ratio of the output voltages from the transformer. The capacities C_{a1} and C_{a2} between the screen and the first active conductors merely shunt the transformer windings; the capacities C_{b1} and C_{b2} between the screens and the second active conductors act as shunts across the detector, so even if the capacities to the screens are large, none of them affects the balance condition.

16·2·4   *Frequency-dependent balance conditions*

The bridges discussed above, when used for simple comparison of two inductances or two capacitances, have balance conditions which are independent of frequency. There is therefore no need to use a source of known frequency, nor even of a single frequency.

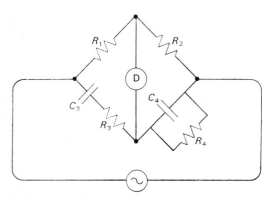

Figure 49.   Wien bridge.

Many other circuits, however, have frequency-dependent balance conditions; a useful example of these, the Wien bridge, may actually be used for measuring frequency. The circuit is shown in figure 49, with two arms pure resistances, one having a capacity and resistance in series, and the fourth a capacity and resistance in parallel. The algebra goes as follows:

$$Re_1 = R_1 \quad ; \quad Im_1 = 0$$

$$Re_2 = R_2 \quad ; \quad Im_2 = 0$$

$$Re_3 = R_3 \quad ; \quad Im_3 = -\frac{1}{\omega C_3}$$

$$Re_4 = \frac{R_4}{(1 + \omega^2 C_4^2 R_4^2)} \quad ; \quad Im_4 = \frac{-\omega C_4 R_4^2}{(1 + \omega^2 C_4^2 R_4^2)}$$

With these values equation (16.2) gives:

$$\frac{R_1 R_4}{(1 + \omega^2 C_4^2 R_4^2)} = R_2 R_3 \tag{16.5}$$

and equation (**16.3**) gives:

$$\frac{\omega C_4 R_4^2 R_1}{(1 + \omega^2 C_4^2 R_4^2)} = \frac{R_2}{\omega C_3} \tag{16.6}$$

If equation (**16.6**) is divided by equation (**16.5**), and multiplied by $\dfrac{\omega}{C_4 R_4}$ we extract from these an expression for ω:

$$\omega^2 = \frac{1}{R_3 R_4 C_3 C_4} \tag{16.7}$$

Substituting this back into equation (**16.5**), a frequency-independent second condition for balance is obtained:

$$\frac{R_1}{R_2} = \frac{R_3}{R_4} + \frac{C_3}{C_4} \tag{16.8}$$

This condition can be satisfied automatically if we vary R_4 with R_3, and C_4 with C_3. For example, we may fix R_1 equal to $2R_2$, controlling R_4 and R_3 by ganged switches, so that:

$$R_4 = R_3 = R;$$

equation (**16.12**) then requires:

$$C_4 = C_3 = C$$

This may be achieved by using two ganged variable condensers, which are set to give zero, or minimum, deflection. The frequency is then obtained from a simplified version of equation (**16.7**), namely:

$$\omega = \frac{1}{CR} \tag{16.9}$$

Arms 3 and 4 of this circuit have some of the properties of a resonant circuit; indeed if the output of an amplifier is connected to arms 3 and 4, in place of the source in figure 49, and the voltage across arm 4 is used as input to the same amplifier, sinusoidal oscillations occur with angular frequency:

$$\omega = \frac{1}{\sqrt{(R_3 R_4 C_3 C_4)}} \tag{16.10}$$

If, among a collection of electronic apparatus, you find a sine-wave generator which appears to contain no self-inductance, you may well find that it is based on the principle of the Wien bridge.

While the Wien bridge could be used for measuring a frequency, in practice a frequency would usually be measured with a cathode-ray oscilloscope or a counter-timer. The Wien bridge is interesting as a non-inductive circuit with the property that it can fix, or measure, a frequency.

16·3 Measurement of power

In a d.c. circuit the power consumed in a load may be obtained by measuring the voltage across it, and the current through it; in an a.c. circuit, however, the power dissipated in a load is not generally obtained by multiplying the voltage across it by the current through it. If we use the r.m.s. values of current and voltage, this does give the correct result for a resistive load; but for loads containing inductive or capacitive components, the power-factor cos δ must also be included. From an experimental point of view, this would not be a good way to measure a power.

A direct measurement of power in an a.c. circuit may be obtained with the aid of a special instrument containing two coils. These are arranged with one fixed and one moving, so that the instantaneous value of the couple exerted by the one on the other is proportional to the product of the two currents in them at the particular moment. If one of the coils is in series with the load, and the other in parallel with it (with an extra series resistance included if necessary), this product is proportional to the rate of dissipation of energy in the load at the moment in question. With a hairspring to allow the movable coil to take up an equilibrium displacement proportional to the time-average of the couple on it, the instrument will measure the average rate of dissipation of energy, which is the power. An instrument which does this is called a wattmeter.

16·4 Measurement of energy

16·4·1 *Cost of electric energy*

The cost of supplying electric power may be divided into two elements:

(i) The cost of the fuel which must be used to generate it. This is effectively proportional to the amount of energy consumed.

(ii) The cost of building and maintaining the generating equipment and the distribution system. This depends more on maximum demand than on total demand.

Another complication is the nature of the load. It is more expensive to supply a given power to an inductive load than to supply the same power to a resistive load. This is partly because the currents are larger by a factor sec δ, and partly because the balance of currents between the parts of a three-phase circuit is upset when one of them is feeding an inductive load. This unbalance may force some components and cables to be increased in size by a factor more serious than sec δ.

Despite these complications electricity supply authorities in most countries compromise by charging domestic consumers purely according to energy consumed. The distinction between fuel costs and capital costs is reflected to some extent in the systems which allow for differences in the rate or time at which the energy is consumed. Examples of such systems are those in which the charge per unit of energy is:

(a) higher for the first so many units per quarter-year, but thereafter at a lower rate;

(b) levied only on energy consumed at a rate faster than a preset limit. This system is appropriate when fuel costs are zero, as in the Norwegian hydro-electric system; capital costs are covered by a fixed charge which is related to the limit below which no charge per unit energy is made;

(c) specially low for circuits connected only during off-peak periods, in which capital equipment would otherwise be under-used. The 'off-peak periods' may be specified by time, as in Great Britain, or by a radio-frequency signal sent along the supply wires, as in Switzerland.

Industrial consumers are sometimes charged according to maximum demand, instead of, or as well as, according to energy consumed.

2 Measurement of energy

Despite all this variety of considerations and methods, it remains clear that there is a widespread need to measure the total electrical energy consumed in a particular circuit during a given time. Most meters for doing this depend on the following principle.

As in the wattmeter, we have a series coi' and a parallel coil, but both are fixed and the moving part is a metal disc pivoted near the coils. The operation of the meter depends on eddy currents induced in this disc. The parallel coil acts as a nearly-pure self-inductance across the source of voltage; the current through it is therefore 90° out of phase with the voltage and of magnitude proportional to it. This current causes induced currents in the disc which are in turn 90° out of phase with it; the induced currents are therefore in phase with the source voltage, and proportional to it.

Now the current in the series coil exerts a force on the disc by virtue of the eddy currents flowing in the disc. This force is proportional to the instantaneous product of the eddy current and the current in the series coil. It is therefore proportional to the instantaneous power, the product of source voltage and load current at the given moment.

This force is in a direction tending to rotate the disc about its axis. Instead of having a hairspring to produce a restoring couple proportional to displacement, we allow the disc to turn freely, but limit its angular velocity by making it turn between the poles of a permanent magnet. This magnet sets up another pattern of eddy currents, with which it interacts to produce a dragging couple proportional to the angular velocity. The equilibrium state of such a system is a state of constant angular velocity proportional to the mean force exerted by the coils, and hence to the mean value of the power in the load circuit. If the power remains constant for a given time, the number of revolutions made by the disc is proportional to the time-integral of the power, which is the energy that we want to measure. The revolutions are counted by a train of gears, driving pointers or dials which may be graduated in kilowatt-hours.

Appendix A
Units

During the development of electromagnetic theory many systems of units have been used. Each has had its own particular merits and demerits, being convenient for parts of the subject and less so for others. As an intellectual exercise, the comparison of different systems of electromagnetic units has some value, but it is now generally felt that the habit of working in a single system may, by limiting avoidable complications in the elementary stages, have advantages which outweigh the demerits of the chosen system in certain branches of the subject.

The system of units most generally adopted is the SI (Système International) and it has been used throughout the text of this book. It is based on a system previously known by the initials RMKSA (rationalised metre kilogram second ampere), of which the last four specify units of length, mass, time and current while "rationalised" referred to the choice of which equations should contain a factor 4π.

However, SI units are not the best for clear thought about advanced electromagnetic theory, and it is necessary for the student to be able to use advanced text books based on other systems. Therefore, to ease the task of the student wishing to use one of these texts for more advanced study, we shall provide some information about the most important alternative system of units, namely the Gaussian cgs system.

Gaussian cgs units

In this system the primary physical units are the centimetre (cm), gramme and second, so that the unit of force is the dyne ($= 10^{-5}$ newton), and that of energy the erg ($= 10^{-7}$ joule). The system is not "rationalised", i.e. there is no factor 4π in the equation for the force between two charges; there are no dimensional constants μ_0 and ε_0, and there is no distinction between **H** and **B** or **E** and **D**, in free space. The feature distinguishing the Gaussian from other cgs systems is that a single unit of charge is used throughout, and a factor c is included in the equations when necessary to relate different types of effect.

To translate the material of this book into Gaussian cgs units, we define a cgs electrostatic unit (cgs esu) of charge by a modified form of equation **(2.1)**:

$$\text{electrostatic force, in dynes, } = \frac{qQ}{r^2}. \tag{2.1}$$

Here r is the distance between the charges q and Q, in cm, while q and Q are the magnitudes of the charges, in cgs esu. Comparison with the original equation (2.1) shows that this unit of charge is $1/(3 \times 10^9)$ coulomb. (By this we really mean $\dfrac{10}{c}$, but it is easier to remember the factors if the approximation $c = 3 \times 10^{10}$ cm/sec is used).

We do not distinguish between electric field and flux density in free space: that due to a point charge Q is given by

$$\mathbf{E} = \mathbf{D} = \frac{Q}{r^2}\,\mathbf{r}. \tag{2.2}$$

The development of electrostatics then follows lines parallel to those of chapters 2, 9 and 10; near the beginning we define a cgs esu of potential equal to one erg per cgs esu of charge, which is 300 volts ($10^{-8}c$). The main changes thereafter follow simply from removal of the "rationalising" factor 4π, with abandonment of ε_0 and distinction between $\mathbf{D}$ and $\mathbf{E}$ in free space: equations in chapter 2 lose the factor $\dfrac{1}{4\pi\varepsilon_0}$, equation (9.1) does the same, (9.2) loses $\dfrac{1}{4\pi}$, and (9.3) to (9.7) gain a factor 4π on the right-hand side. Equations (9.8) to (9.12) become:

$$\mathbf{D} = \mathbf{E} + 4\pi\,\mathbf{P} \tag{9.8}$$

$$\mathbf{P} = \chi_e\,\mathbf{E} \tag{9.9}$$

$$\mathbf{D} = \varepsilon\,\mathbf{E} \tag{9.10}$$

$$\varepsilon = 1 + 4\pi\chi_e \tag{9.11}$$

$$\mathbf{E} = 4\pi\sigma/\varepsilon \tag{9.12}$$

The cgs esu of capacity is $\dfrac{1}{9 \times 10^{11}}$ farad $\left(= \dfrac{10^9}{c^2} \right)$, and the expressions for capacities in these units are:

Isolated sphere: a, the radius in cm $\hfill$ (10.1)

Concentric spheres: $\dfrac{\varepsilon r_1 r_2}{r_2 - r_1}$ $\hfill$ (10.3)

Coaxial cylinders: $\varepsilon / \left\{ 2\log_e\left(\dfrac{r_2}{r_1}\right) \right\}$ per unit length $\hfill$ (10.4)

Parallel plates: $\dfrac{\varepsilon A}{4\pi d}$ $\hfill$ (10.5)

Here, and in the rest of this appendix, equations are given the same label as the corresponding equation in the main text.

In the Gaussian cgs system, the magnetic force between two moving charge is calculated by methods identical to those of chapter 4 and Appendix C modified only by omission of the factor $4\pi\varepsilon_0$, right up to the point when the force is given by equation (C.21), which is equivalent to equations (4.10) and (4.15 combined as:

$$\mathbf{F}_m = q\mathbf{v} \times \left(\frac{1}{c^2}\mathbf{u} \times \mathbf{E} \right).$$

$\left(\dfrac{1}{c}\mathbf{u}\times\mathbf{E}\right)$, where $\mathbf{E}$ is the electrostatic field of the source-charge in cgs esu, is called the magnetic field. This magnetic field is in oersteds, and we shall call it $\mathbf{B}^*$, as a private notation to distinguish it from $\mathbf{H}$ and $\mathbf{B}$ of the main text. Comparison of this definition with equation (4.10), allowing for the different units of $\mathbf{E}$, shows that

$\qquad \mathbf{B}^*$ (in oersteds) $= 10^4\,\mathbf{B}$ (in weber m^{-2}),

whence

$\qquad \mathbf{B}^*$ (in oersteds) $= 4\pi\times 10^{-3}\,\mathbf{H}$ (in ampere m^{-1}).

In this system, we make no distinction between magnetic field and flux density in free space, so that the units oersted $\left(=\dfrac{1000}{4\pi}\text{ ampere/m}\right)$ and gauss ($= 10^{-4}$ weber/m^2) are interchangeable.

With $\mathbf{B}^*$ incorporating one of the factors $\dfrac{1}{c}$, the other must appear in the expression relating $\mathbf{F}_m$ to $\mathbf{B}^*$, which becomes

$$\mathbf{F}_m = \frac{q}{c}\,\mathbf{v}\times\mathbf{B}^* \qquad (4.15)$$

so that the total Lorentz force is now

$$\mathbf{F} = q\left(\mathbf{E}+\frac{1}{c}\,\mathbf{v}\times\mathbf{B}^*\right) \qquad (4.16)$$

This leads to an expression for the force on a current element in a magnetic field:

$$\mathbf{F} = \frac{I}{c}\,\delta\mathbf{s}\times\mathbf{B}^* \qquad (4.21)$$

Calling the magnetic field of a moving charge Q

$$\mathbf{B}^* = \frac{1}{c}\,\mathbf{u}\times\mathbf{E} \qquad (4.10)$$

$$= \frac{Q}{cr^2}\,\mathbf{u}\times\mathbf{r} \qquad (4.14)$$

leads us to give the magnetic field of a current element as

$$\mathbf{B}^* = \frac{I}{cr^2}\,\delta\mathbf{s}\times\mathbf{r} \qquad (4.19)$$

i.e $\qquad \mathbf{B}^* = \dfrac{I}{c}\,\dfrac{\delta s\,\sin\theta}{r^2} \qquad (4.20)$

From here, we calculate the magnetic fields of wires and coils as in chapter 5, with $\dfrac{I}{4\pi}$ replaced by $\dfrac{I}{c}$. Some important results are:

Long straight wire:

$$\mathbf{B}^* = \frac{2I}{ca} \qquad (5.4)$$

At centre of a circular coil of n turns:

$$\mathbf{B}^* = \frac{2\pi nI}{ac} \qquad (5.6)$$

Inside a long solenoid:

$$\mathbf{B}^* = 4\pi \, m \, \frac{I}{c} \tag{5.9}$$

The line integral of magnetic field around a wire becomes

$$\oint \mathbf{B}^* \cdot \mathbf{ds} = \frac{4\pi I}{c}, \tag{5.10}$$

and Maxwell's equation for curl $\mathbf{B}^*$ is:

$$\text{curl } \mathbf{B}^* = \frac{1}{c}\left(4\pi j + \frac{d\mathbf{E}}{dt}\right) \tag{5.16}$$

The forces between currents are calculated from the above expressions, with results equivalent to those of chapter 6 with the rationalising factor $\frac{1}{4\pi}$ removed, and μ_0 replaced by $\frac{1}{c^2}$; the force per unit length between parallel straight wires becomes

$$F = \frac{2I_1 I_2}{c^2 a} \tag{6.3}$$

and the force on a current element ds_2 due to a current element ds_1 is:

$$\mathbf{F}_2 = \frac{I_1 I_2}{c^2 r^2} \, \mathbf{ds}_2 \times (\mathbf{ds} \times \mathbf{r}) \tag{6.4}$$

The calculations of induced emfs and electric fields follow the argument of chapter 7, with a new definition of magnetic flux: for this we use AB^* in free space and $A(1 + 4\pi\chi_m)B^*$ in material of magnetic susceptibility χ_m. In terms of B^* the couple on a coil is $\frac{I}{c} A B^* \sin\theta$. The rate of doing work on a rotating coil is therefore

$$\frac{I}{c} A B^* \sin\theta \frac{d\theta}{dt} = -\frac{I}{c} \frac{dN}{dt}$$

where $N = A B^* \cos\theta$ is the newly defined flux through the coil. The induced emf must therefore be written:

$$V = -\frac{1}{c} \frac{dN}{dt}. \tag{7.1}$$

Following this, Maxwell's equation for curl $\mathbf{E}$ becomes:

$$\text{curl } \mathbf{E} = -\frac{1}{c} \frac{d\mathbf{B}^*}{dt}. \tag{7.25}$$

Self-inductance is defined as induced emf per unit rate of change of current; we thus have a factor $\frac{1}{c}$ in the Gaussian cgs version of equation (7.1) and N contains a factor $\frac{1}{c}$ instead of μ_0; self-inductances may therefore be obtained from the equations of chapter 7, with $\frac{4\pi}{c^2}$ inserted instead of μ_0. (The 4π comes from "un rationalisation"). But L is not the same as the flux per unit current; it is $\dfrac{N}{cI}$.

Lastly, when we consider the equivalence of coil and magnet, we define a magnetic moment m in terms of the couple per unit B^*. Then, with the couple on a coil given by $\dfrac{I}{c} A B^*$, equivalence requires

$$m = \frac{I}{c} A. \tag{8.3}$$

Conversion factors

In the table below, we give some useful conversion factors; each entry is the size of the cgs unit, expressed in SI units. It is also the number by which a quantity in cgs units must be multiplied to obtain the same quantity in SI units.

Table of conversion factors

Quantity	SI unit	Gaussian cgs unit
Charge	coulomb	$= \dfrac{1}{(3 \times 10^9)} C$
Current	ampere	$= \dfrac{1}{(3 \times 10^9)} A$
Potential	volt	$= 300\ V$
Electric field	volt metre^{-1}	$= 3 \times 10^4\ V\ m^{-1}$
Capacity	farad	$= \dfrac{1}{(9 \times 10^{11})} F$
Resistance	ohm	$= 9 \times 10^{11} \Omega$
Inductance	henry	$= 9 \times 10^{11} h$
Magnetic field	ampere metre^{-1}	oersted $= \dfrac{1000}{4\pi} A\ m^{-1}$
Magnetic flux density	weber metre^{-2} $=$ Tesla	gauss $= 10^{-4}$ Tesla
Magnetic flux	weber	gauss cm$^2 = 10^{-8}$ weber

For maximum accuracy, replace each factor 3 by 2·99793, the velocity of light in units of 10^8m/sec. Take 9 as $(2\cdot99793)^2$.

Note also: $\mu_0 = 4\pi \times 10^{-7}$ henry metre^{-1}

$$\epsilon_0 = \frac{1}{(4\pi \times 9 \times 10^9)}\ \text{farad metre}^{-1}$$
$$= 8.85416 \times 10^{-12}\ \text{farad metre}^{-1}$$

Appendix A

Appendix B
Vector methods

Properties of vectors – vector addition

When two quantities which have direction as well as magnitude are to be added together, it is necessary to specify the type of addition required; this depends on the purpose of the calculation. If a man walks three miles northwards and then four miles eastwards, we may want to know how far he has walked. This will be just seven miles, obtained by arithmetical addition of three miles to four miles; this is the distance which would have to be divided by his velocity to find the time taken for the journey. But if we want to know how far he is from his starting-point, we must take into account the angle between the two legs of his journey: in this case, with the angle 90°, the answer is five miles (i.e. $\sqrt{[3^2 + 4^2]}$). The full description of his position with respect to his starting-point is five miles in a direction 37° north of east'; this is called the vector sum of the two vectors which made up the two straight parts of his journey.

Conversely, if we wish to specify the man's position as five miles from the starting-point, in a direction 37° north of east, we may state that his position relative to the starting-point has components three miles north and four miles east. This says nothing about the route he actually followed. If we extend his movement by sending him up a 30-foot pole whose base is at this position, we may say that his displacement has components three miles north, four miles east and 30 feet upwards, or that his position has coordinates three miles north, four miles east and 30 feet up. Neither of these statements indicates whether he reached the top of the pole by climbing up it after walking to the base, or by moving horizontally after rising 30 feet, or by going up an inclined path. We may think one route is more probable than the others, but in general a route needs more words to specify it than does a relative position; the latter is specified by a vector, which may be described either by its magnitude and direction, or by its components. If a route, or part of a route, is straight, it may be considered as a vector; but if it is curved, it needs a more elaborate description.

Vectors of position are the easiest to visualize, so much so that they are often used, in diagrams, to illustrate the properties of other types of vector. These other types include vectors of velocity, momentum, force and acceleration. All of these can be resolved into components, and added, by the same rules that apply to vectors of position. For example a velocity of five miles per hour north-east may be said to have components $\dfrac{5}{\sqrt{2}}$ north and $\dfrac{5}{\sqrt{2}}$ east, without implying that these components have any separate physical significance. Or the force exerted by a bicycle tyre on the ground may be described as the vector sum of a vertical force of 400 newtons due to the rider's weight, and a horizontal force of 40 newtons due to centrifugal force as he turns a corner; in these circumstances, the total, or resultant, force is 402 newtons in a direction inclined at 5·7° from vertical.

The notation which we have taken as standard is that in which a symbol in ordinary type denotes the magnitude of a vector, e.g. p for momentum, and the symbol in bold type, e.g. $\mathbf{p}$, denotes the vector itself; the magnitudes of the components along a set of axes x, y and z are indicated by subscripts, as in p_x, p_y and p_z. If the x-component of $\mathbf{p}$ is required as a vector in its own right, it must be represented by $\mathbf{p}_x$, or by $p_x\mathbf{i}$, where $\mathbf{i}$ is a unit vector in the direction of the x-axis.

As we have already indicated without discussion, the addition of two vectors may be described as a geometrical process, the resultant being the third side of the triangle which has as its other two sides the two vectors which are to be added. But in order to avoid this geometrical description, we may say that the resultant of two vectors is a vector, each of whose components is the sum of the corresponding components of the two vectors. Thus by adding the velocity $\mathbf{v}$, with components v_x, v_y, v_z, to another velocity $\mathbf{u}$, with components u_x, u_y, u_z, the vector sum or resultant is obtained as a velocity:

$$\mathbf{V} = \mathbf{v} + \mathbf{u},$$

with components:
$$V_x = v_x + u_x$$
$$V_y = v_y + u_y$$

and:
$$V_z = v_z + u_z$$

Scalar product

A scalar is a quantity which has a magnitude, but which is not associated with any particular direction. Its magnitude should not be affected by the choice of the directions of any axes that may be used for describing vector quantities. A specially interesting scalar quantity may be formed from two vectors, in the following manner.

Let us define the scalar product of vector $\mathbf{a}$ with vector $\mathbf{b}$ as the magnitude of $\mathbf{a}$ multiplied by the projection of $\mathbf{b}$ in the direction of $\mathbf{a}$.

If $\mathbf{b}$ has components $\mathbf{b}_x$, $\mathbf{b}_y$ and $\mathbf{b}_z$, the projection of $\mathbf{b}$ in the direction of $\mathbf{a}$ is the sum of the projections of its three components in this direction. This can be seen for the two-dimensional case in figure 50, where the projection of $\mathbf{b}$ in the direction of $\mathbf{a}$ is OQ, while those of its components are OP and PQ.

The projection of $\mathbf{b}_x$ in the direction of $\mathbf{a}$ is just its magnitude b_x multiplied by the cosine of the angle between $\mathbf{a}$ and the x-axis.

This cosine may be expressed as the ratio $\dfrac{a_x}{a}$, where a and a_x are the magnitudes of $\mathbf{a}$ and its x-component.

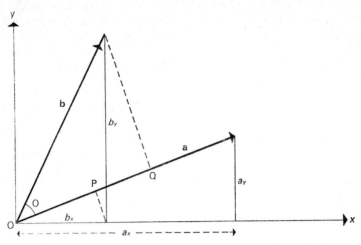

Figure 50. Two-dimensional diagram to illustrate scalar product.

The projection of b_x in the direction of a is therefore $\dfrac{b_x a_x}{a}$.

Similar arguments apply to the y- and z-components of b, so we conclude that the sum of the projections of the three components of b in the direction of a is:

$$\frac{b_x a_x + b_y a_y + b_z a_z}{a}$$

The scalar product of a with b was defined as a multiplied by the above quantity, which gives us:

$$b_x a_x + b_y a_y + b_z a_z$$

Similarly, the scalar product of b with a, defined as the magnitude of b multiplied by the projection of a in the direction of b, becomes equal to:

$$a_x b_x + a_y b_y + a_z b_z$$

which is the same as the scalar product of a with b. Indeed both these quantities have to be equal to:

$$ab \cos \theta$$

where θ is the angle between the vectors a and b.

The scalar product of two vectors a and b may be defined in any of the ways which have been shown above to be equivalent. Since neither $ab \cos \theta$, nor the definition from which we started, are affected by any choice of axes, it is clear that the adjective scalar in the name scalar product is merited. It is, however, interesting to note that we have proved the scalar nature of the quantity:

$$a_x b_x + a_y b_y + a_z b_z$$

which was formed by multiplication of the components of the two vectors **a** and **b**. It was not obvious that this quantity would be independent of our choice of axes.

It is useful to note that the scalar product of two mutually perpendicular vectors is zero, whereas on the other hand, the scalar product of **a** with itself is just the square of its length.

Vector product

It has just been shown how two vectors may be multiplied in such a way as to give a scalar quantity, their scalar product; they may alternatively be multiplied to give a vector quantity, which we shall now discuss under the name of their vector product.

The vector product of two vectors **a** and **b** may be defined as a vector perpendicular to the plane of **a** and **b**, of magnitude equal to the area of the parallelogram formed by **a** and **b**. The vector so defined will have a component in any

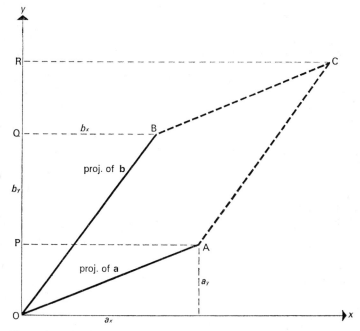

Figure 51.   Vector product $a \times b$: z-component obtained by projection on to the x–y plane.

direction equal to the area of the parallelogram projected on to the plane perpendicular to that direction. For example, the vector product of **a** with **b** will have a z-component equal to the area of the projected parallelogram OACB in figure 51. This area is given by:

Area OACB = Area OAP + Area ACRP − Area CBQR − Area OBQ

$$= \tfrac{1}{2}a_x a_y + b_y(a_x + \tfrac{1}{2}b_x) - a_y(b_x + \tfrac{1}{2}a_x) - \tfrac{1}{2}b_y b_x$$

$$= a_x b_y - a_y b_x$$

This is therefore the expression for the z-component of the vector product of **a** with **b**, in terms of the components of **a** and **b**.

We may project the parallelogram formed by **a** and **b** similarly on to the yz and zx planes, where x, y and z are a set of right-handed axes, as illustrated in figure 14. These give the x- and y-components of the vector product, as follows:

$$x\text{-component} = a_y b_z - a_z b_y$$

$$y\text{-component} = a_z b_x - a_x b_z$$

There are two symbols in common use for vector product, $\times$ and $\wedge$; thus the vector product of a with b, discussed above, is written either as **a** × **b** or as **a**∧**b**. If we write:

$$\mathbf{c} = \mathbf{a} \times \mathbf{b},$$

the components of **c** are given by:

$$c_x = a_y b_z - a_z b_y$$

$$c_y = a_z b_x - a_x b_z$$

$$c_z = a_x b_y - a_y b_x$$

It will be noticed that there is an element of convention in this definition **a** × **b** has been given a positive component along an axis about which a rotation of **a** to **b** involves a *clockwise* rotation (through less than 180°), as we look from the origin along the axis. **b** × **a** would be equal to − **a** × **b**, and which is which follows from our decision to use a right-handed set of axes. Two examples of directions of vector products are shown in figure 7.

When two vectors **a** and **b** are at right-angles, **a** × **b** has magnitude ab in a direction at right-angles to both **a** and **b**. On the other hand, when two vectors are parallel, their vector product is zero in magnitude, and has no defined direction.

The curl of a vector

In Chapter 5, the idea of the curl of a vector was developed, and the label curl **H** was applied to the limit, for a small closed path, of the vector normal to the plane of the path, whose magnitude is the line integral of **H** round the path divided by the area enclosed by the path.

This may be taken as the definition of the curl of a general vector **A**; its component in any direction has magnitude:

$$\frac{\oint \mathbf{A} \cdot \mathbf{ds}}{\delta S}$$

where the line integral is taken round a small closed path in the plane perpendicular to that direction. δS is the area enclosed by the path.

In terms of the components of **A**, and of **ds**, the above expression is:

$$\frac{\oint (\;{}^{_1}{}_x dx + A_y dy + A_z dz)}{\delta S}$$

In order to calculate the z-component of curl **A**, we have to choose a path in the x–y plane, in which $\oint A_z dz$ will make no contribution. The easiest path is the rectangular one shown in figure 52, for which δS is equal to $\delta x \, \delta y$.

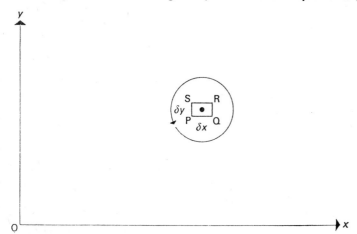

Figure 52. Rectangular path for evaluation of z-component of a curl.

It gives, as the z-component of curl **A**, the limit of:

$$\frac{1}{\delta x \, \delta y} \oint_{PQRS} (A_x dx + A_y dy)$$

$$= \frac{1}{\delta x \, \delta y} \left[\int_P^Q A_x dx + \int_Q^R A_y dy + \int_R^S A_x dx + \int_S^P A_y dy \right]$$

$$= \frac{1}{\delta x \, \delta y} \left[\int_P^Q A_x dx - \int_S^R A_x dx - \int_P^S A_y dy + \int_Q^R A_y dy \right]$$

$$= \frac{1}{\delta x \, \delta y} \left[\left(A_x - \frac{\delta y}{2} \frac{\partial A_x}{\partial y} \right) \delta x - \left(A_x + \frac{\delta y}{2} \frac{\partial A_x}{\partial y} \right) \delta x - \right.$$

$$\left. - \left(A_y - \frac{\delta x}{2} \frac{\partial A_y}{\partial x} \right) \delta y + \left(A_y + \frac{\delta x}{2} \frac{\partial A_y}{\partial x} \right) \delta y \right]$$

$$= \left(\frac{\partial A_y}{\partial x} - \frac{\partial A_x}{\partial y} \right)$$

where in the last two lines A_x, A_y, $\dfrac{\partial A_x}{\partial y}$ and $\dfrac{\partial A_y}{\partial x}$ are all evaluated at the centre of the path. If the path is shrunk to the limit at which it becomes a point, the z component of curl $\mathbf{A}$, defined as above, is therefore equal to the value of $\left(\dfrac{\partial A_y}{\partial x} - \dfrac{\partial A_x}{\partial y}\right)$ at that point.

Corresponding arguments give expressions for the x- and y-components of curl $\mathbf{A}$. The results may be summed up in the equations:

If $\quad \mathbf{B} = \text{curl } \mathbf{A}$

$$B_x = \frac{\partial A_z}{\partial y} - \frac{\partial A_y}{\partial z}$$

$$B_y = \frac{\partial A_x}{\partial z} - \frac{\partial A_z}{\partial x}$$

$$B_z = \frac{\partial A_y}{\partial x} - \frac{\partial A_x}{\partial y}$$

Stokes's Theorem

The definition of curl $\mathbf{A}$ as the limiting value of the vector whose component normal to a small area δS is $\dfrac{\oint \mathbf{A} \cdot \mathbf{ds}}{\delta S}$, shows us that the value of (curl $\mathbf{A}$) . δS for a small area is equal to that of the line integral $\oint \mathbf{A} \cdot \mathbf{ds}$ round it. Extending this to a larger area, over which curl $\mathbf{A}$ may not be uniform, and which need not itself be planar, we see that the surface integral of curl $\mathbf{A}$ round the perimeter of the surface, i.e.

$$\int \text{curl } \mathbf{A} \cdot \mathbf{dS} = \oint \mathbf{A} \cdot \mathbf{ds}.$$

The divergence of a vector

In Chapter 8, we discussed the total outward flux of a vector over a closed surface. For a small closed surface, it was shown that the total outward flux per unit volume enclosed by the surface was, for a general vector $\mathbf{A}$:

$$\frac{\partial A_x}{\partial x} + \frac{\partial A_y}{\partial y} + \frac{\partial A_z}{\partial z}$$

This was called the divergence of A or div $\mathbf{A}$.

If we go back to a larger closed surface, we see that the total outward flux of $\mathbf{A}$ over the surface must equal to the volume integral of div $\mathbf{A}$ over the enclosed volume, i.e.

$$\oint \mathbf{A} \cdot \mathbf{dS} = \int dv \text{ div } \mathbf{A}.$$

This is a mathematical law of general validity and is known as Gauss's Law.

The gradient of a scalar

It was mentioned in Chapter 2, that the electric field may be written as the gradient of a potential. This is an important example of a widely-used technique. A scalar quantity has a numerical value which may vary from place to place, but has no directional property itself; examples are density, pressure, charge density, gravitational potential and electric potential. We obtain a vector, from such a scalar, by differentiating it. The x-component of the gradient of a scalar ϕ is the partial differential $\dfrac{\partial \phi}{\partial x}$, while the y- and z-components are $\dfrac{\partial \phi}{\partial y}$ and $\dfrac{\partial \phi}{\partial z}$. It is sometimes written grad ϕ.

The operator ∇

In the preceding sections, we have considered the partial differential coefficients of the three components of a vector $\mathbf{A}$. It is sometimes convenient to consider these as the result of allowing a vector operator ∇ (called nabla, or sometimes del) to act on $\mathbf{A}$: $\dfrac{\partial}{\partial x}$ is considered as the x-component of ∇, $\dfrac{\partial}{\partial y}$ as its y-component, and $\dfrac{\partial}{\partial z}$ as its z-component.

If ∇ acts on $\mathbf{A}$ in such a manner that it forms a scalar product, we get:

$$\nabla . \mathbf{A} = \nabla_x A_x + \nabla_y A_y + \nabla_z A_z$$

However, these are not products, since ∇_x, ∇_y, and ∇_z are operators; but they make up the terms $\dfrac{\partial A_x}{\partial x}$, $\dfrac{\partial A_y}{\partial y}$, $\dfrac{\partial A_z}{\partial z}$ which occur in the divergence of $\mathbf{A}$. For this reason, the symbol $\nabla . \mathbf{A}$ is sometimes used for div $\mathbf{A}$.

Examining the components of curl $\mathbf{A}$ in the same way, we find that they resemble the components of a vector product written $\nabla \times \mathbf{A}$. If curl $\mathbf{A}$ is written $\nabla \times \mathbf{A}$, this has the additional usefulness of reminding us that the sign convention for curl $\mathbf{A}$ is related to that for the vector product.

In this notation, the gradient of a scalar ϕ may be written as $\nabla \phi$, which we must remember is a vector.

Appendix C
The force between two moving charges, in special relativity

Note: This appendix supplies a more rigorous and detailed version of the arguments presented in Chapter 4, with proper attention to all three components of the velocities. It is considerably more advanced than the rest of the book.

4-vectors

4-vectors are four-dimensional vectors, in most cases with three components closely related to the components of an ordinary three-dimensional vector. For consistency with the ordinary vector notation, a bold-type letter, e.g. A, is used to represent a three-dimensional vector, while bold-type underlined, e.g. $\underline{A}$, indicates the corresponding 4-vector. The fourth component is a quantity chosen so that the sum of the squares of all four components is invariant for some physical or algebraic reason. By 'invariant' we mean that it appears the same when measured in different frames of reference moving with constant relative velocity. This includes frames, with relative velocity zero, related to each other by rotation of the spatial axes, which will leave unaltered the sum of the squares of the three spatial components. It also includes frames in relative motion along one of the spatial axes, in which case 4-vectors in the two frames are related by the simplest form of Lorentz transformation; this involves only the one spatial component and the fourth component, as follows:

Suppose that in the frame Σ, a 4-vector $\underline{A}$ has components A_x, A_y, A_z, A_4; in another frame Σ', moving with velocity u along the x-axis of Σ, the 4-vector would be called $\underline{A}'$, and would seem to have components A_x', A_y', A_z', A_4', given by the equations:

$$\left.\begin{array}{lll} A_x' = \gamma_0 A_x & & + i\gamma_0\beta A_4 \\ A_y' = & A_y & \\ A_z' = & A_z & \\ A_4' = -i\gamma_0\beta A_x & & + \gamma_0 A_4 \end{array}\right\} \quad \begin{array}{l} \text{with } \beta = \dfrac{u}{c} \\[2mm] \text{and } \gamma_0 = (1-\beta^2)^{-\frac{1}{2}} \end{array} \qquad \text{(C.1)}$$

These equations are conveniently written as one matrix equation:

$$A' = TA \qquad\qquad\qquad\qquad \text{(C.2)}$$

Here A' and A represent the components of the two 4-vectors, arranged as column matrices $\begin{bmatrix} A_x' \\ A_y' \\ A_z' \\ A_4' \end{bmatrix}$ and $\begin{bmatrix} A_x \\ A_y \\ A_z \\ A_4 \end{bmatrix}$ while T is a 4×4 square matrix containing the coefficients γ_0 and $i\gamma_0\beta$, as follows:

$$T = \begin{bmatrix} \gamma_0 & 0 & 0 & i\gamma_0\beta \\ 0 & 1 & 0 & 0 \\ 0 & 0 & 1 & 0 \\ -i\gamma_0\beta & 0 & 0 & \gamma_0 \end{bmatrix}$$

The rules of matrix multiplication ensure that the product TA is a column matrix identical with the right-hand side of equation (C.1).

Without writing out the equations in full, we shall simply state that the reverse transformation, from Σ' to Σ, is described by a matrix:

$$R = \begin{bmatrix} \gamma_0 & 0 & 0 & -i\gamma_0\beta \\ 0 & 1 & 0 & 0 \\ 0 & 0 & 1 & 0 \\ i\gamma_0\beta & 0 & 0 & \gamma_0 \end{bmatrix}$$

It may be checked by ordinary algebra that 4-vectors satisfying equation (C.2) also satisfy:

$$A = RA' \tag{C.3}$$

It may also be checked that the transformations T and R satisfy the condition of leaving unchanged the sum of the squares of the four components of A and A'.

The 4-vectors which will be required most often are the coordinate 4-vector, of form:

$$\underline{X} = \begin{bmatrix} x \\ y \\ z \\ ict \end{bmatrix}$$

and the momentum-energy 4-vector, or 4-momentum, of form:

$$\underline{p} = \begin{bmatrix} p_x \\ p_y \\ p_z \\ \dfrac{iW}{c} \end{bmatrix}$$

where p_x, p_y, p_z are the components of ordinary 3-dimensional momentum $\mathbf{p}$, and W is energy. We shall also need the 4-vectors which are obtained by differentiating these with respect to proper time. The proper time τ is the time measured in the rest-frame of the particle under consideration, and it is related to the time t measured in another frame by:

$$t = \gamma\tau,$$

where:

$$\gamma = \left(1 - \frac{v^2}{c^2}\right)^{-\frac{1}{2}}$$

and v is the relative velocity of the frames.

$\dfrac{d\mathbf{X}}{d\tau}$ is called the four-velocity, and is equal to:

$$\mathbf{v} = \begin{bmatrix} \dfrac{dx}{d\tau} \\ \dfrac{dy}{d\tau} \\ \dfrac{dz}{d\tau} \\ \dfrac{icdt}{d\tau} \end{bmatrix} = \gamma \begin{bmatrix} v_x \\ v_y \\ v_z \\ ic \end{bmatrix} \tag{C.4}$$

where $v_x = \dfrac{dx}{dt}$, the x-component of ordinary 3-dimensional velocity $\mathbf{v}$; v_y and v_z are the y and z components of $\mathbf{v}$. The four-velocity is odd in having a constant fourth component, but it may be seen easily that, with the factor γ multiplying all four components, this ensures an invariant sum of squares of the four components.

Differentiation of the four-momentum yields the four-force $\mathbf{F}$. This has its first three components γ times larger than the components of the ordinary three-dimensional force $\mathbf{f}$, defined as rate of change of momentum. The steps are as follows:

$$\mathbf{F} = \dfrac{d\mathbf{p}}{d\tau} = \gamma \begin{bmatrix} \dfrac{dp_x}{dt} \\ \dfrac{dp_y}{dt} \\ \dfrac{dp_z}{dt} \\ \dfrac{i}{c}\dfrac{dW}{dt} \end{bmatrix} = \gamma \begin{bmatrix} f_x \\ f_y \\ f_z \\ \dfrac{i}{c}\mathbf{f}\cdot\mathbf{v} \end{bmatrix} \tag{C.5}$$

In the fourth component, we have replaced the rate of change of energy, $\dfrac{dW}{dt}$, by the equivalent rate of doing work by the force $\mathbf{f}$, as it moves its point of application with a velocity $\mathbf{v}$; this rate of doing work is the scalar product $\mathbf{f}\cdot\mathbf{v}$.

Electric charges

Let us now consider a source-charge Q, at rest at the origin of the frame Σ', i.e. having dashed coordinates:

$$x' = y' = z' = 0.$$

In the frame Σ, it will be moving with velocity u along the x-axis, and will have coordinates:

$$x = ut = \beta ct$$

$$y = 0$$

$$z = 0,$$

if the origins of the two frames are arranged to coincide at time $t = 0$.

We also need a test-charge q, moving through the frame Σ with velocity $\mathbf{v}$, so that at time $t = 0$ it has coordinates x_0, y_0, z_0; at time t it will have coordinates:

$$x = x_0 + v_x t$$

$$y = y_0 + v_y t$$

$$z = z_0 + v_z t$$

In the frame Σ, it will have a 4-velocity:

$$\underline{\mathbf{v}} = \gamma \begin{bmatrix} v_x \\ v_y \\ v_z \\ ic \end{bmatrix} \tag{C.6}$$

where $\gamma = \left(1 - \dfrac{v^2}{c^2}\right)^{-\frac{1}{2}}$ and $v = (v_x^2 + v_y^2 + v_x^2)^{\frac{1}{2}}$

In the frame Σ', q will have coordinates given by:

$$\underline{X'} = \begin{bmatrix} x' \\ y' \\ z' \\ ict' \end{bmatrix} = T \begin{bmatrix} x \\ y \\ z \\ ict \end{bmatrix} = \begin{bmatrix} \gamma_0(x - ut) \\ y \\ z \\ i\gamma_0(ct - \beta x) \end{bmatrix}$$

When $t = 0$, these become $\underline{X'_0} = \begin{bmatrix} \gamma_0 x_0 \\ y_0 \\ z_0 \\ -i\beta\gamma_0 x_0 \end{bmatrix}$ \tag{C.7}

In the frame Σ', q will have a velocity $\mathbf{v}'$, which may be calculated by the transformation relating the four-velocities $\underline{\mathbf{v}}$ and $\underline{\mathbf{v}}'$:

$$\underline{\mathbf{v}}' = T\underline{\mathbf{v}} = \gamma \begin{bmatrix} \gamma_0(v_x - \beta c) \\ v_y \\ v_z \\ i\gamma_0(c - i\beta v_x) \end{bmatrix} \tag{C.8}$$

The 3-dimensional velocity $\mathbf{v}'$ is obtained by dividing the first three components of $\underline{\mathbf{v}}'$ by γ'. $\gamma' = \left(1 - \dfrac{v'^2}{c^2}\right)^{-\frac{1}{2}}$ cannot be evaluated for the moment, since it involves v', the magnitude of $\mathbf{v'}$. However, it is sufficient to specify that:

$$\mathbf{v'} = \frac{\gamma'}{\gamma'}\begin{bmatrix} \gamma_0(v_x - \beta c) \\ v_y \\ v_z \end{bmatrix} \qquad \text{(C.9}$$

Force between the electric charges

In the frame Σ', which is the rest-frame of the source-charge Q, the only fiel
due to Q is an electrostatic field $\mathbf{E'}$. If charge is invariant, the test-charge q
whatever its velocity, will experience a 3-dimensional force, measured in Σ
equal to:

$$\mathbf{f'} = q\mathbf{E'} = \frac{qQ}{4\pi\varepsilon_0 r'^3}\,\mathbf{r'} \qquad \text{(C.10}$$

This is taken as axiomatic, and it leads us to state that the 4-force on q, measure
in Σ', is:

$$\underline{\mathbf{F'}} = \gamma'\frac{qQ}{4\pi\varepsilon_0 r'^3}\begin{bmatrix} x' \\ y' \\ z' \\ \dfrac{i}{c}\mathbf{r'}.\mathbf{v'} \end{bmatrix} \qquad \text{(C.11}$$

following equation (C.5).
  Returning to the laboratory frame Σ, we find that q experiences a four-forc
given by:

$$\underline{\mathbf{F}} = R\,\underline{\mathbf{F'}} = \gamma\frac{qQ}{4\pi\varepsilon_0 r'^3}\begin{bmatrix} \gamma_0\left(x' + \dfrac{\beta}{c}\mathbf{r'}.\mathbf{v'}\right) \\ y' \\ z' \\ i\gamma_0\left(\beta x' + \dfrac{1}{c}\mathbf{r'}.\mathbf{v'}\right) \end{bmatrix} \qquad \text{(C.12}$$

The 3-dimensional force $\mathbf{f}$ on q, measured in Σ, is obtained by dividing the fir
three components of $\underline{\mathbf{F}}$ by γ, which has already been defined as the gamma
factor corresponding to the velocity of q in Σ. This gives:

$$\mathbf{f} = \frac{\gamma'}{\gamma}\frac{qQ}{4\pi\varepsilon_0 r'^3}\begin{bmatrix} \gamma_0\left(x' + \dfrac{\beta}{c}\mathbf{r'}.\mathbf{v'}\right) \\ y' \\ z' \end{bmatrix} \qquad \text{(C.13}$$

We may substitute in this expression the value of $\mathbf{v}'$ given by (C.9), obtaining:

$$\mathbf{f} = \frac{\gamma'}{\gamma} \frac{qQ}{4\pi\varepsilon_0 r'^3} \begin{bmatrix} \gamma_0\left\{x'\left(1 - \beta^2\gamma_0\frac{\gamma}{\gamma'}\right) + \frac{\gamma}{\gamma'}\frac{\beta}{c}(x'\gamma_0 v_x + y'v_y + z'v_z)\right\} \\ y' \\ z' \end{bmatrix}$$

(C.14)

From equation (C.7), when $t = 0$:

$$x' = \gamma_0 x_0, \ y' = y_0, \ z' = z_0, \ \text{whence} \ \underline{\mathbf{r}}' = (\gamma_0^2 x_0^2 + y_0^2 + z_0^2)^{\frac{1}{2}}.$$

These values may be substituted into equation (C.14), to give:

$$\mathbf{f} = \frac{qQ}{4\pi\varepsilon_0}(\gamma_0^2 x_0^2 + y_0^2 + z_0^2)^{-\frac{3}{2}} \begin{bmatrix} \gamma_0^3 x_0\left(\dfrac{\gamma'}{\gamma\gamma_0} - \beta^2\right) + \gamma_0\dfrac{\beta}{c}(\gamma_0^2 x_0 v_x + y_0 v_y + z_0 v_z) \\ \dfrac{\gamma'}{\gamma}y_0 \\ \dfrac{\gamma'}{\gamma}z_0 \end{bmatrix}$$

(C.15)

For a test-particle at rest in Σ, we must put $v_x = v_y = v_z = 0$, $\gamma = 1$ and $\gamma' = \gamma_0$; these substitutions, with $1 - \beta^2 = \gamma_0^{-2}$, reduce (C.15) to the much simpler form:

$$\mathbf{f}_{\text{stat}} = \frac{qQ}{4\pi\varepsilon_0}(\gamma_0^2 x_0^2 + y_0^2 + z_0^2)^{-\frac{3}{2}} \begin{bmatrix} \gamma_0 x_0 \\ \gamma_0 y_0 \\ \gamma_0 z_0 \end{bmatrix}$$

(C.16)

This force $\mathbf{f}_{\text{stat}}$ is attributed to the electrostatic field $\mathbf{E}$ in the frame Σ, and must be put equal to $q\mathbf{E}$. $\mathbf{E}$ is therefore given by:

$$\mathbf{E} = \frac{Q}{4\pi\varepsilon_0}(\gamma_0^2 x_0^2 + y_0^2 + z_0^2)^{-\frac{3}{2}} \begin{bmatrix} \gamma_0 x_0 \\ \gamma_0 y_0 \\ \gamma_0 z_0 \end{bmatrix}$$

(C.17)

It is interesting to compare $\mathbf{E}$ with $\mathbf{E}'$, the electrostatic field at the same place and time, measured in the rest-frame of the source-charge. $\mathbf{E}'$, for $t = 0$, is given by substituting:

$$\mathbf{r}' = \begin{bmatrix} \gamma_0 x_0 \\ y_0 \\ z_0 \end{bmatrix} \ \text{in (C.10), obtaining:}$$

$$\mathbf{E}' = \frac{Q}{4\pi\varepsilon_0}(\gamma_0^2 x_0^2 + y_0^2 + z_0^2)^{-\frac{3}{2}} \begin{bmatrix} \gamma_0 x_0 \\ y_0 \\ z_0 \end{bmatrix}$$

(C.18)

Comparison of (C.18) and (C.17) shows that $\mathbf{E}'$ and $\mathbf{E}$ have identical components along the direction of motion of the source (the x-direction); but the z and y

components differ by a factor γ_0. These transformations of electrostatic field may be summarized by:

$$E_x = E'_x$$
$$E_y = \gamma_0 E'_y$$
$$E_z = \gamma_0 E'_z \qquad\qquad\qquad\qquad (C.19)$$

The magnetic force

The difference between $\mathbf{f}$, the force on a moving test-charge, and $\mathbf{f}_{stat}$, the force on a stationary test-charge, is attributed to magnetic effects. The magnetic force $\mathbf{f}_{mag}$ is obtained by subtracting (C.16) from (C.15): with some algebraic manipulation of the x-component, this gives:

$$\mathbf{f}_{mag} = \frac{qQ}{4\pi\,\varepsilon_0(\gamma_0^2 x_0^2 + y_0^2 + z_0^2)^{\frac{3}{2}}}\gamma_0 \begin{bmatrix} \gamma_0^2 x_0\left(\dfrac{\gamma'}{\gamma\gamma_0}-1\right) + \dfrac{\beta}{c}(\gamma_0^2 x_0 v_x + y_0 v_y + z_0 v_z) \\[2ex] \left(\dfrac{\gamma'}{\gamma\gamma_0}-1\right)y_0 \\[2ex] \left(\dfrac{\gamma'}{\gamma\gamma_0}-1\right)z_0 \end{bmatrix}$$

$$(C.20)$$

In order to simplify this expression for $\mathbf{f}_{mag}$, we must evaluate the factor $\left(\dfrac{\gamma'}{\gamma\gamma_0}-1\right)$; provided that the velocities are small compared with c, we may write:

$$\gamma = 1 + \frac{v^2}{2c^2}, \text{ where } v \text{ is the velocity of } q \text{ in the frame } \Sigma,$$

$$\gamma_0 = 1 + \frac{u^2}{2c^2}, \text{ where } u \text{ is the velocity of } Q \text{ in the frame } \Sigma,$$

$$\gamma' = 1 + \frac{v'^2}{2c^2}, \text{ where } v' \text{ is the velocity of } q \text{ in the frame } \Sigma';$$

this is the velocity of q relative to Q, which in this approximation may be written:

$$\mathbf{v}' = \mathbf{v} - \mathbf{u}$$

This vector equality implies that the magnitudes of the velocities are related by:

$$v'^2 = v^2 + u^2 - 2vu \cos \theta,$$

where θ is the angle between $\mathbf{v}$ and $\mathbf{u}$. But $v \cos \theta = v_x$, since $\mathbf{u}$ is along the x-axis, therefore:

$$v'^2 = v^2 + u^2 - 2uv_x,$$

and

$$\left(\frac{\gamma'}{\gamma\gamma_0}-1\right) = 1+\frac{v'^2}{2c^2}-\frac{v^2}{2c^2}-\frac{u^2}{2c^2}-1$$

$$= \frac{1}{2c^2}(v'^2-u^2-v^2)$$

$$= -\frac{uv_x}{c^2}$$

Substituting this in equation (C.20), and replacing β by $\dfrac{u}{c}$ in the second part of the x-component, we find that the $x_0 v_x$ terms cancel, leaving:

$$f_{mag} = \frac{qQ}{4\pi\varepsilon_0(\gamma_0^2 x_0^2+y_0^2+z_0^2)^{\frac{3}{2}}}\ \gamma_0\frac{u}{c^2}\begin{bmatrix} y_0 v_y+z_0 v_z \\ -y_0 v_x \\ -z_0 v_x \end{bmatrix} \tag{C.21}$$

The r.h.s. of (C.21) has the form of a vector product of $\mathbf{v}$ with something whose x-component is zero; we may therefore write equation (C.21) as:

$$f_{mag} = q\mathbf{v}\times\mathbf{B} \tag{C.22}$$

where:

$$\mathbf{B} = \frac{Q}{4\pi\varepsilon_0 c^2}(\gamma_0^2 x_0^2+y_0^2+z_0^2)^{-\frac{3}{2}}\ \gamma_0 u\begin{bmatrix} 0 \\ -z_0 \\ y_0 \end{bmatrix} \tag{C.23}$$

The r.h.s. of (C.23), having zero x-component, may be expressed as the vector product of something with $\mathbf{u}$, which is along the x-axis. Comparison of equations (C.17) and (C.23) shows that the something is $\dfrac{1}{c^2}\mathbf{E}$. Equation (C.23) is therefore equivalent to:

$$\mathbf{B} = \frac{1}{c^2}\mathbf{u}\times\mathbf{E} \tag{C.24}$$

This expression must hold for any $\mathbf{u}$, since any general case may be reduced to the one considered above, by choosing new axes such that x is along $\mathbf{u}$.

We have now reached the conventional description of the magnetic force between two moving charged particles, as it was quoted in Chapter 4. Equations (C.22) and (C.24) are identical, respectively, with equations (4.15) and (4.10). For their development and discussion, the reader is therefore referred to Chapters 4, 5 and 6.

Further reading

References have been included in the text to indicate sources of further information on particular points. In making the following suggestions about more general reading, nothing is implied about the books which have escaped mention. There are plenty of excellent books on all aspects of electromagnetism, and selection of a few titles must involve some arbitrary and subjective elements.

At roughly the same level as the present text, and supplying details which may usefully supplement it, are E. M. Purcell, *Electricity and Magnetism*, volume 2 of the Berkeley Physics Course, McGraw Hill, 1963, and B. I. Bleaney and B. Bleaney, *Electricity and Magnetism*, Oxford, 1957. For help in the fields which their titles indicate, the student may turn to D. E. Rutherford, *Vector Methods*, Oliver & Boyd, 1943, and to W. G. V. Rosser, *Introduction to the Theory of Relativity*, Butterworth, 1964.

Covering much of the material of the present text in rather more advanced language, and extending it to cover electromagnetic radiation, are J. C. Slater and N. H. Frank, *Electromagnetism*, McGraw Hill, 1947, J. R. Reitz and F. J. Milford, *Foundations of Electromagnetic Theory*, Addison-Wesley, 1960, F. N. H. Robinson, *Electromagnetism*, Oxford, 1974.

For more advanced reading on circuit theory, one might select G. Newstead, *General Circuit Theory*, Methuen, 1959, and J. Millman and H. Taub, *Pulse and Digital Circuits*, McGraw Hill, 1956. On electrical techniques there is much useful information in F. E. Terman and J. M. Pettit, *Electrical and Electronic Measurements*, McGraw Hill, 1952, H. V. Malmstadt and C. G. Enke, *Electronics for Scientists*, Benjamin, 1963, L. Strauss, *Wave Generation and Shaping*, McGraw Hill, 1960, and P. D. Ankrum, *Semiconductor Electronics*, Prentice-Hall, 1971.

Lastly as an advanced text on fundamentals, useful in the later years of an honours degree course, a good choice would be J. D. Jackson, *Classical Electrodynamics*, Wiley, 1962.

Index